Mouse Hematology

A Laboratory Manual

ALSO FROM COLD SPRING HARBOR LABORATORY PRESS

RELATED LABORATORY MANUALS

Basic Methods in Microscopy: Protocols and Concepts from Cells, A Laboratory Manual
Imaging in Neuroscience and Development: A Laboratory Manual
Live Cell Imaging: A Laboratory Manual, Second Edition
Manipulating the Mouse Embryo: A Laboratory Manual, Third Edition

OTHER RELATED TITLES

Mouse Phenotypes: A Handbook of Mutation Analysis
At the Bench: A Laboratory Navigator, Updated Edition
At the Helm: A Laboratory Navigator
Lab Dynamics: Management Skills for Scientists
Lab Math: A Handbook of Measurements, Calculations, and Other Quantitative Skills for Use at the Bench
Lab Ref: A Handbook of Recipes, Reagents, and Other Reference Tools for Use at the Bench

RELATED WEBSITE

Cold Spring Harbor Protocols
www.cshprotocols.org

MOUSE HEMATOLOGY

A LABORATORY MANUAL

MICHAEL P. MCGARRY
CHERYL A. PROTHEROE
JAMES J. LEE

Mayo Clinic Arizona

CSH PRESS
COLD SPRING HARBOR LABORATORY PRESS
Cold Spring Harbor, New York • www.cshlpress.com

Mouse Hematology

A Laboratory Manual

Published exclusively by Cold Spring Harbor Laboratory Press
Printed in the United States of America

Publisher	John Inglis
Acquisition Editor	David Crotty
Director of Development, Marketing, & Sales	Jan Argentine
Developmental Editor	Maria Smit
Project Coordinator	Inez Sialiano
Production Editor, Book	Rena Steuer
Production Editor, Poster	Kathleen Bubbeo
Production Editor, DVD	Mala Mazzullo
Desktop Editor	Lauren Heller
Production Manager	Denise Weiss
Book Marketing Manager	Ingrid Benirschke
Sales Account Managers	Jane Carter and Elizabeth Powers
Cover Designer	Ed Atkeson

Front cover artwork: Brush smear of wild-type (C57BL/6J) mouse bone marrow (400x).

Library of Congress Cataloging-in-Publication Data

McGarry, Michael P., 1942-
Mouse hematology : a laboratory manual / Michael P. McGarry, Cheryl A. Protheroe, James J. Lee.
p. cm.
Includes bibliographical references and index.
ISBN 978-0-87969-885-0 (cloth : alk. paper) -- ISBN 978-0-87969-886-7 (pbk. : alk. paper)
1. Hematology, Experimental--Laboratory manuals. 2. Mice as laboratory animals--Laboratory manuals. 3. Veterinary hematology--Laboratory manuals. I. Protheroe, Cheryl A., 1963- II. Lee, James J., 1958- III. Title.

QP91.M385 2010
612.40078--dc22

2009038914

10 9 8 7 6 5 4 3 2 1

All Cold Spring Harbor Laboratory Press publications may be ordered directly from Cold Spring Harbor Laboratory Press, 500 Sunnyside Blvd., Woodbury, New York 11797-2924. Phone: 1-800-843-4388 in the Continental U.S. and Canada. All other locations: (516) 422-4100. FAX: (516) 422-4097. E-mail: cshpress@cshl.edu. For a complete catalog of all Cold Spring Harbor Laboratory Press publications, visit our visit our website at http://www.cshlpress.com/.

In memory of Jim Kern — *M.P.M., C.A.P., J.J.E.*

False views, if supported by some evidence, do little harm,
for everyone takes a salutary pleasure in proving their falseness.

– Charles Darwin

Contents

VIDEOS ON ACCOMPANYING DVD

POSTER

A quick reference guide to mouse peripheral blood cells, including brief descriptions of each lineage, total cellularities, and hematological/physiological parameters of interest.

Preface

RECENT ADVANCES IN FLOW CYTOMETRIC TECHNOLOGIES, molecular diagnostics, and the sophistication of immunophenotyping of hematopoietic cell types highlight a significant diagnostic shift with provocative implications for the assessment of mouse models of human diseases. In particular, the use of clusters of differentiation markers, receptors, and other entities expressed on cell surfaces; molecules present in the cytosol (and/or associated with subcellular structures); and the expression patterns of specific transcription factors is now common. All have but displaced the classical identification of blood cell lineages in Romanowsky dye-stained preparations of peripheral blood, bone marrow, and other lymphohematopoietic tissues of laboratory mice by light microscopic examination. As a result, hematopoietic cell lineages and cell types are now defined by a complex and extensive expression repertoire that only remotely resembles the nomenclature traditionally used to identify the cells that comprise these compartments. We recently made this observation and commented on its significance as it related to the interpretation of data derived from mouse models of allergic diseases (Lee and McGarry 2007). Furthermore, the use of gene expression profiles in fresh versus cultured cells has further led to a separation of contemporary classifications from traditional nomenclature.

In general, there is widespread use of mouse models in laboratories in which the basic biology, anatomy, physiology, and histology/cytology of the laboratory mouse have not been part of preparative training. We believe that many research laboratories may benefit from a concise review of the blood and blood-forming tissues of the laboratory mouse as well as some of the conventional techniques for the preparation of blood/marrow samples and their examination by light microscopy. Unfortunately, although very useful texts contain comprehensive descriptions of human hematopoiesis (e.g., Custer and Hayhoe 1974; Tavassoli and Yoffey 1983; Koury et al. 2009), there is a relative paucity of published work containing a comprehensive morphological description of the blood and blood-forming tissues of the laboratory mouse. The classical text from The Jackson Laboratory on mouse biology (Russell and Bernstein 1975) contains a valuable chapter on mouse hematology with typical values for selected parameters, but little is available regarding definitive cellular morphologies. Veterinary hematology textbooks similarly have some valuable descriptive sections on comparative hematology in mammals, including the mouse (Schalm et al. 1975). Their value in part attests to the comparability of hematopoiesis and the formed elements of blood across species; although differences exist, the similarities of both form and process are striking. However, once again, the specific assessments of mouse hematological parameters, including cell differentials, are poorly described in these textbooks.

Our goal here is to present standards and procedures for the examination of blood and blood-forming tissues of the laboratory mouse, using male and female C57BL/6J and BALB/cJ mice as source material. Although we have had experience with the hematology of a wide variety of strains of inbred mice as well as random-bred, hybrid, and genetically manipulated mice, it is our objective to present broad principles and strategies for the effective morphological identification of the cellular elements of blood and blood-forming tissues. Subtle differences may exist among some mouse strains; however, these differences do not preclude the identification of lineages. The techniques and approaches discussed should make it possible to avoid confusion that may be encountered in many, if not most, experimental settings. That is, the described methodologies will allow for the morphological examination of blood, bone marrow, and/or hematopoietic tissues in research protocols and, in turn, will provide a greater understanding of mouse models of human disease.

We thank the enormous contributions of the following associates at Mayo Clinic Arizona whose help made this project possible: Randy Raish and Bill Bedrava (Mayo Clinic Media Support Services) provided us with videographic support that was truly unbelievable—simply outstanding! Nikki Boruff and Marv Ruona (Mayo Clinic Visual Communications) tirelessly worked on the figures in this book, allowing us to present clear and wonderfully detailed examples of mouse blood cells—their efforts were key to the success of this book. We also acknowledge Drs. Nancy A. Lee, Elizabeth Jacobson, Sergei Ochkur, and the staff of the Lee Laboratories for the fantastic job they did in critically reading various drafts of this manuscript, as well as the helpful technical assistance of Katie O'Neill who continually made sure that mice were available as the project proceeded. We also thank Linda Mardel and Shirley ("Charlie") Kern for their help with the essential administrative functions of our lab and the production of this work. To be sure, this book would have never seen the light of day if it were not for the dedicated and wonderfully supportive staff at Cold Spring Harbor Laboratory Press, including Inez Sialiano (project coordinator) and developmental editor, Maria Smit, without whom we would surely have lost our minds, production editor Rena Steuer, desktop editor Lauren Heller, and production manager Denise Weiss. Finally, we want to extend a special note of gratitude to an old friend and colleague, David Crotty (executive editor, CSH Protocols), whose encouragement and enthusiasm propelled us to begin and, more importantly, complete this project.

MICHAEL P. MCGARRY
CHERYL A. PROTHEROE
JAMES J. LEE

1 Collection of Peripheral Blood

ACROSS ALL COMMON STRAINS OF MICE, the volume of blood in a mature, otherwise healthy animal of either gender ranges from 5% to 7% of body weight. Thus, for a 25-g mouse, the blood volume is ~1.25–1.75 mL. However, only 60%–75% of this can be sampled by whole-blood collection techniques, including terminal cardiocentesis under anesthesia (Hoff 2000). For blood films and cell counts, a much smaller volume is necessary (see Chapters 2 and 3). Normally, to make a blood film, a small sample (e.g., a drop) is most easily obtained from the tail vasculature of a mouse that has been physically restrained either manually or in a device.

Before obtaining the sample, it is necessary to assess the general health of the subject animal(s) and the purpose for the collection. If serial samples are to be obtained and the animal is immune deficient, some consideration of the procedures must be given (e.g., ensuring sterile techniques may be necessary). Similarly, if the sample is to be used for culture, sterility will be necessary. Extra precautions must be exercised for any animal that is thrombocytopenic or simply a "bleeder" as a result of any number of causes (e.g., congenital mutations, cytotoxic pharmacotherapeutics, or other transgenic or experimental interventions); otherwise, it may exsanguinate during the collection. Conversely, some animals may have low blood volumes and/or be prone to rapid clotting and thus may not yield sufficient blood for slide preparation.

Technical Tip: It is not possible to list here all of the pathophysiologic variables that may impact the quality or quantity of the blood from a particular research mouse. However, be assured that the state of hydration, infectious pathogens, genetic manipulation, and other conditions may, and likely will, impact the blood and circulatory system, which in turn will affect the sample collected.

Several techniques are available for accessing the vasculature either to obtain a drop of blood on a slide or to collect 10–100 μL using a glass capillary tube (Morton et al. 1993; Hoff 2000; Golde et al. 2005; Christensen et al. 2009). The collection site is usually determined by the accessibility of the vasculature, the frequency of sampling, and whether the sample must be sterile for other applications. Some prefer a toe, the saphenous vein, or other sites; however, we prefer the tail for several reasons: The tail vasculature is easily accessed without chemical restraint, the process is relatively quick because no additional steps (e.g., depiliation) are necessary, the target veins are easily visualized, postsampling observations are simple, the procedure can easily be done by one person, and serial sampling via the tail is less traumatic than with many other methods. Methods for obtaining blood samples from the tail are demonstrated in Videos 1 and 2 of the accompanying DVD.

Collection from a lateral tail vein is described in Video 1 (Venous Access for Blood Film Using Lateral Tail Vein). For serial samples, initially select a distal site of the lateral tail vein for collection. Make subsequent blood preparations from cuts more proximal to the base of the tail. It is also possible to perform the serial sampling by moving proximally down both sides of the tail, alternating between the two lateral tail veins. This pattern of multiple blood collections is important because after incision, the healing of the veins may affect blood flow. Moreover, the vessels will remain quite visible proximal (but not distal) to sites of previous samples. An alternative location, and our site of choice, is the most distal tip of the tail (see Video 2, Venous Access for Blood Film Using Tip of Tail). After a site is chosen, make a clean, simple 1-mm cut. Discard the initial droplet that forms so that the collected sample is not contaminated by interstitial cells. Never force blood from the cut by squeezing the tail, although some gentle pressure is acceptable. Forced collection may result in contamination with endothelial cells or it may skew the composition in other ways. Sufficient volumes for either films or cell counts (or both) may be obtained from the tail vasculature.

Technical Tip: Because the site, volume, frequency, and method of collection are matters for review and approval by the local Institutional Animal Care and Use Committee (IACUC), consult the veterinary staff before the submission of a protocol and/or its implementation.

Although blood collection methods are virtually the same for both adults and younger animals, special consideration must be given for collecting blood from neonatal and other preweaned mice. The blood of younger animals generally clots more rapidly, and the cut surface may result in the rejection of the pup by the dam when it is returned to its cage. The risk of rejection is greater for younger pups. Similar to adults, the tail and toe are the most common sites for blood collection in pups. After obtaining the desired sample from a pup, stop the bleeding by compression, and wipe all traces of blood from the pup using a water-soaked gauze. In addition, roll the pup in the soiled bedding of the cage before it is returned to the cage. Even still, the dam may reject the pup, ignore it, or at worst, cannibalize it. In this event, some of the techniques used to encourage acceptance of pups by lactating dams for foster nursing may be helpful.

2 Counting Red Blood Cells, Platelets, and Viable Nucleated White Blood Cells

ALTHOUGH SOME RESEARCH LABS HAVE ACCESS to automated blood-cell-counting technologies, many do not and thus they must rely on simple methods to assess the various parameters for the formed elements of the blood, including the nucleated white blood cells (WBCs or leukocytes), the anucleate hemoglobin-containing red blood cells (RBCs), and platelets. An approximate indication of the cellular composition of blood may be made by centrifuging a capillary tube that is partially filled with peripheral blood as described in Protocol 1. The collection of blood for centrifugation is generally performed with capillary tubes treated with an anticoagulant (e.g., heparin) to prevent platelet activation and the clotting cascade.

Did you know? White cells are called white because when peripheral blood is centrifuged, it forms a small "white" band at the interface between the packed red cells and the clear plasma. This band of cells, called the "buffy coat," is also reflective of the number of nucleated white cells in the peripheral circulation.

Counting the number of red or white cells per unit volume of peripheral blood may be accomplished by any number of techniques (see http://www.irvingcrowley.com/cls/fund.htm), some of which use a special pipette (an RBC pipette or a WBC pipette). The use of these pipettes is now limited because they require the development of some skill; they will not be described here. In Protocols 2, 3, and 4, we describe more common and easily performed methods of counting RBCs, platelets, and WBCs, respectively. Because RBCs are anucleate, viability is not as critical as with nucleated white cells that may be isolated for in vitro tests or in vivo transfer protocols.

Technical Tip: It is always preferable to record all data (including animal and cohort identifiers, date/time, laboratory notebook/page, and IACUC protocol number[s]) associated with a given blood sample (i.e., cell counts and differentials) in a single venue. Tables 1 and 2 provide convenient yet practical templates for recording such information.

TABLE 1. Worksheet Template for Cell Counts

Investigator ______________________________ Date/Time______________________________

IACUC Protocol Number ___________________ Laboratory Book No./Page No.________________

Experiment/Protocol Title ____

Animal group and/or cohort	Mouse tag no.	Hematocrit	Total cell count (cells/mm^3)		
			RBC (x10^6)	Platelets (x10^5)	WBC (x10^3)

TABLE 2. Worksheet Template for Differential Analyses

Investigator ____________ Date/Time ____________

IACUC Protocol Number ____________ Laboratory Book No./Page No. ____________

Experiment/Protocol Title ____________

Animal group and/or cohort	Mouse tag no.	Tissue[a]	Total no. leukocytes assessed	WBC (i.e., leukocyte) differential assessment									
				Lymphocyte (Lym)		Monocyte (Mo)		Neutrophil (Neu)		Eosinophil (Eos)		Other	
				Total	%	Total	%	Total	%	Total	%	Total	%

[a] For example, blood, bone marrow, spleen, extramedullary hematopoietic site.

Protocol 1

Determination of Percent Red Cell Volume (Hematocrit)

The hematocrit (in Greek, hemo means blood and crit means to separate) is the percent of packed red cells in a unit volume of peripheral blood. This measure is easily obtained with the use of a microhematocrit centrifuge as described here. Once spun, the RBCs pack at the bottom of the tube. The clear volume, composed of either serum or platelet-rich plasma, occupies the top of the tube. The leukocytes form a band at the interface.

MATERIALS

CAUTION: See Appendix for proper handling of materials marked with <!>.

Reagents/Animals

Mouse from which a blood sample is to be collected

Equipment

Clay trays to plug the bottom end of the capillary tube

Heparinized capillary tubes

> If commercial preheparinized capillary tubes are not available, heparinize glass capillary tubes (which may be obtained from a variety of sources) by submersion in 1× PBS containing 20 units/mL heparin <!> for 1 h. Once treated, drain and air-dry the tubes before use. Prepared tubes may be stored indefinitely in closed containers at room temperature.

Microhematocrit centrifuge

> This is a single-purpose device; there are two rotors, one for normal capillary tubes and one for microcapillary tubes. For an alternative to a microhematocrit centrifuge, see Discussion.

Microhematocrit reader

METHOD

1. Collect blood into a heparinized capillary tube by capillary action, filling each tube approximately three-quarters of the way (Morton et al. 1993; Hoff 2000; Golde et al. 2005; Christensen et al. 2009). Make sure to maintain accurate association between the tube and the animal from which the sample was taken.

2. Use one finger to hold the top end of the filled tube and place it vertically in a clay tray to form a sealed plug at the other end of the tube. Be careful not to break the tube.
3. Place the tubes into the rotor of the microhematocrit centrifuge with the clay end to the periphery of the rotor, making sure to associate the sample number with the slot number in the rotor. Cover the rotor, tighten it down, lock the lid, and then centrifuge (at fixed speed) for 8–10 min at room temperature.
4. Use a microhematocrit reader to obtain the percentage of RBCs in the tube, according to the manufacturer's instructions.

   The hematocrit for a normal adult mouse will generally fall between 39% and 44%. Anemia is variably defined, but it should be considered if the hematocrit (i.e., the packed red cell volume) falls below 28%. Polycythemia, the result of overproduction of red cells, may be defined as a hematocrit of >55%–60%, depending on the lab. Platelet counts (Protocol 3) are helpful to determine the etiology of various bleeding disorders. The thickness of the buffy coat (i.e., nucleated peripheral WBCs) is normally <2% of the separated volume. The plasma should be clear (see Troubleshooting).

TROUBLESHOOTING

Problem (Step 4): Plasma color is red, indicating hemolysis.

Solution: Hemolysis may result from fragile RBCs or an autoimmune hemolytic anemia. In most cases, this phenomenon may be reflective of a number of disorders or the consequences of contaminated equipment and/or procedures associated with soaps or organic solvents (e.g., alcohols), to name a few. If extensive, hemolysis will result in an underestimate of the percent red cells and the procedure must be repeated to help assess the cause.

DISCUSSION

If a microhematocrit centrifuge is not available, several alternative strategies for assessing the hematocrit of experimental mice are possible using equipment found in most laboratory settings. The key to any of these strategies is to ensure that the long axis of the microhematocrit tube is in the same plane as the centrifugal force during centrifugation. Figure 1 shows a solution that we developed using glass conical vials with inserts (Thermo Scientific Reacti-Vials) that fit in the buckets of a Sorvall RT 6000D tabletop centrifuge. In this strategy, the capillary tubes containing blood (and plugged with clay at the bottom) are held vertical by the inserts and the conical shape of the Reacti-Vials, which, in turn, fit snuggly in the

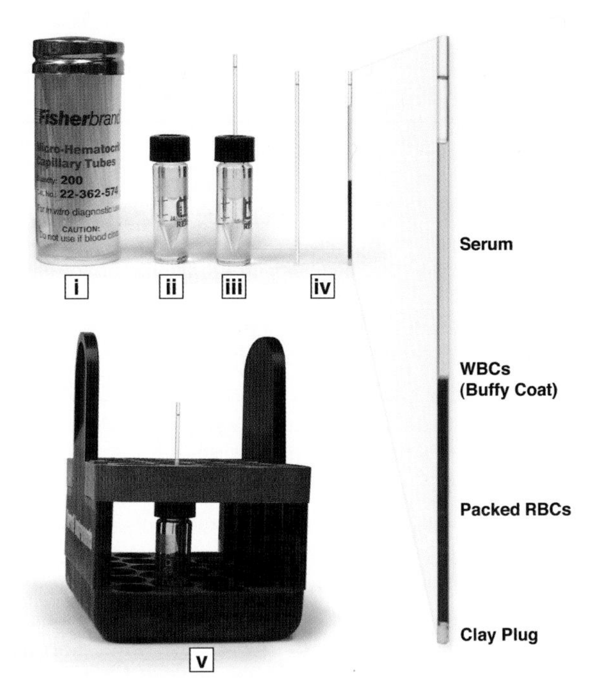

FIGURE 1. Components used to "work around" the need for a microhematocrit centrifuge include (*i*) standard heparinized microhematocrit capillary tubes, (*ii*) Reacti-Vial (Thermo Scientific), (*iii*) a Reacti-Vial with a microhematocrit capillary tube inserted vertically and held in place by the vial's screw-top insert, (*iv*) control balance tube (*left*) relative to a microhematocrit capillary tube that contains a blood sample (*right*), and (*v*) Reacti-Vial sitting snuggly in the "multiple 15-mL tube" bucket used in a Sorvall RT 6000D tabletop centrifuge. Note the fractional separation of blood components from top to bottom in the microhematocrit capillary tube containing a spun blood sample (*iv, right*): clarified serum, WBCs or "buffy coat" layer, and packed RBCs.

buckets of the tabletop centrifuge. The samples are centrifuged (3500 rpm, ~1800*g*) for 5 min at room temperature. The volume of packed RBCs relative to the total blood volume in the spun capillary is measured as a fraction with the aid of a ruler, and the hematocrit (i.e., percent packed RBCs) is determined. It is noteworthy that there is nothing inherently special about this approach to measure the hematocrits of mice. It simply represents one of many solutions to the logistical problem associated with the need for centrifuging these capillary tubes such that the tube's long axis is in the plane of the centrifugal force applied.

Protocol 2

Methodology to Determine a Red Blood Cell Count Using a Hemacytometer

This protocol describes how to assess the number of RBCs in peripheral blood using a hemacytometer.

MATERIALS

Reagents/Animals

Dulbecco's phosphate-buffered saline (PBS; 1×), prechilled to 4°C
Mouse from which a blood sample is to be collected

Equipment

Bucket with ice
Compound light microscope with a 40× objective lens and 10× oculars
Conical tube (15 mL, polystyrene)
Coverslips (glass)
Hemacytometer (Fig. 2)
Heparinized capillary tubes

If commercial preheparinized capillary tubes are not available, heparinize glass capillary tubes (which may be obtained from a variety of sources) by submersion in 1× PBS containing 20 units/mL heparin for 1 h. Once treated, drain and air-dry the tubes before use. Prepared tubes may be stored indefinitely in closed containers at room temperature.

Vortex mixer (*optional*; see Step 1)

METHOD

1. Collect 25 µL of blood in a heparinized capillary tube (Morton et al. 1993; Hoff 2000; Golde et al. 2005; Christensen et al. 2009) and express it into a 15-mL polystyrene conical tube maintained on ice that contains 10 mL of cold 1× PBS. Mix thoroughly using a mechanical vortex mixer or by repeatedly inverting the sealed tube.
2. Assemble the hemacytometer by centering a coverslip onto it and placing the unit on a flat bench surface in preparation for counting the number of cells in a given sample.

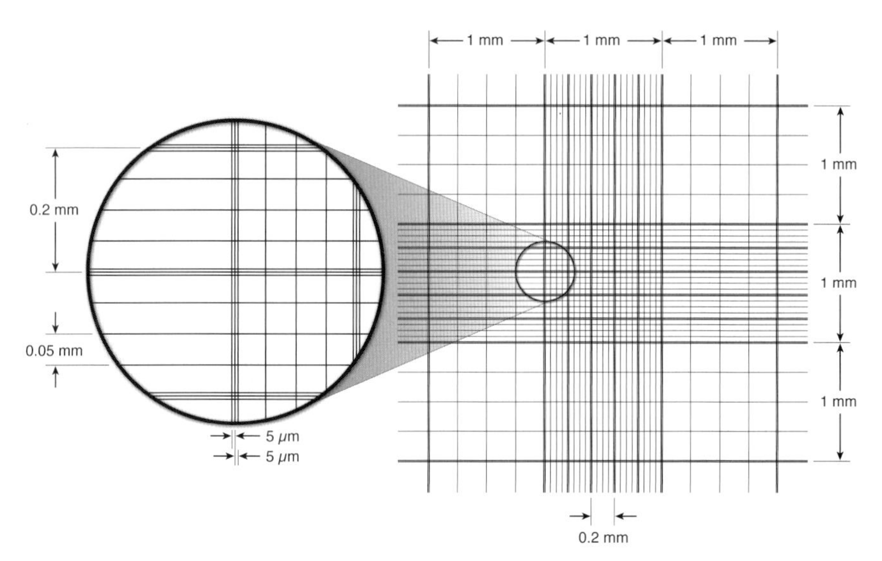

FIGURE 2. Detailed schematic representation of the grid system and the etched surface dimensions of a standard hemacytometer. Hemacytometers are bilaterally symmetrical slides with grids of increasing resolution etched onto the slide's surface. Each half of the hemacytometer contains a "tic-tac-toe"-like grid with four large etched squares (one each at the four corners of the grid), each further divided into 16 smaller etched squares. In addition, the grid also contains a large center square of equal area but with a finer degree of etching. This center square of the grid contains within it 25 smaller etched squares, each containing 16 yet smaller etched squares. Each of these 16 smaller squares corresponds to an area of 1/400 mm^2. Finally, each of the five large squares of the grid covers a surface area of 1 mm^2. The depth of the cell suspension between the coverslip and the slide of a properly loaded hemacytometer is 0.1 mm, permitting the calculation of final cell concentrations as cells/mm^3.

3. Place a small amount (5–10 µL) of the suspension on a clean hemacytometer, sufficient to fill the area over the counting grid under the coverslip but not enough to "float" the coverslip or overrun the scored surface.

   ▸ *See Troubleshooting.*

4. Focus on the large central square of the grid (the square with the finest degree of etching); this large central square is divided into 25 smaller squares. Use the 40x objective lens and 10x oculars of the microscope to count the RBCs in five of these smaller squares—the four corner squares as well as the centermost square. Keep the tally for each square separately. In each count, include RBCs that touch the borders of only two sides of a given square (i.e., top and bottom or left and right). Do not include those touching the borders of the reciprocal two sides.

   We have found that this strategy provides a conveniently accurate way to count the RBCs; i.e., when the RBCs touching all of the square's sides are included, the RBC counts were found to be overestimates. If all RBCs touching the sides of the squares were eliminated from the count, the resulting levels significantly underestimated the true RBC count. Consistency in the counting strategy is what is important—not the particular rules adopted.

▸ *See Troubleshooting.*

5. Calculate the RBC count of the peripheral blood sample in question as follows:

   i. Average the five values from Step 4.

   ▸ *See Troubleshooting.*

   ii. Multiply the average RBC count by 25 (i.e., the total number of boxes in the large central square of the grid) to obtain the total number of RBCs in the central square of the grid.

   iii. The large central square has an area of 1 mm^2 and a depth of 0.1 mm; therefore, multiply the value obtained in Step 5.ii by 10 to generate the number of RBCs/mm^3 in the suspension that was loaded onto the hemacytometer.

   iv. Multiply the concentration of RBCs in the suspension by the initial dilution factor used to generate the suspension (i.e., 1:400, if 25 µL of blood was diluted in 10 mL of 1x PBS in Step 1).

TROUBLESHOOTING

Problem (Step 3): Fluid "floats" the coverslip and/or there are bubbles between the hemacytometer and the coverslip.

Solution: If too much fluid has been applied, removing some of the volume with a towel or other absorbent tissue can skew the distribution of cells over the grid or disproportionately affect cell numbers. Likewise, bubbles will distort cell distribution; never use squares with bubbles for a count because the air displaces fluid volume that normally would contain cells to be included in the count. If either of these occur, it is best to clean the hemacytometer and begin afresh.

Problem (Step 4): Red blood cells appear to be damaged or only membrane "ghosts" seem to be present.

Solution: Lysis from residual soaps or other contaminants can cause damaged (i.e., hemolyzed) samples by leaving residual membrane fragments, referred to as "ghosts" or empty RBCs. In addition, the use of an undiluted concentrated stock of saline (e.g., 10x PBS) will result in shrunken (crenated) red cells. Therefore, always begin with clean dry tubes, pipettes, and other materials, as well as properly balanced and diluted stock diluents.

Problem (Step 4): There are too few (<20) or too many (>120) RBCs per square.

Solution: Given the estimates of normal RBC counts in mice (see the poster, Mouse Peripheral Blood Cells), the dilution steps should be planned to yield counts, on average, of 75–100 RBCs per each of the five (four corner and center) squares indicated in Step 4. If counts are outside of this range, first verify that the dilutions were performed appropriately; if so, adjust the dilution steps to target the suggested range to ensure statistically valid, significant numbers. In the end, the animal may be anemic (few RBCs) or plethoric (many RBCs), either of which will need to be documented.

Problem (Step 5.i): One or more of the counts obtained in Step 4 deviates by more than ~10% from the mean calculated in Step 5.i.

Solution: The RBC numbers of the individual squares are counted and recorded independently to ensure an even distribution of the suspension in the hemacytometer. The individual values within the counted squares of the larger center square of the grid should not deviate by more than ~10% from the mean. If this occurs, disassemble, clean, and reassemble the hemacytometer before reloading the cell suspension for counting.

Protocol 3

Methodology to Determine a Platelet Cell Count Using a Unopette and a Hemacytometer

Platelet counts are best done using commercially available disposable devices that are designed for this purpose. Unopette is a user-friendly and convenient method, combining a simple capillary tube for collecting the sample directly from the cut vein of the animal and a container with a measured amount of diluent with an anticoagulant to prevent platelet activation. A sample is then taken from this reservoir and loaded directly onto the hemacytometer for a platelet count.

MATERIALS

Reagents/Animals

Mouse from which a blood sample is to be collected

Equipment

Compound light microscope with a 40× objective lens and 10× oculars
Coverslips (glass)
Hemacytometer (Fig. 2)
Unopette (Becton Dickinson)
Vortex mixer

METHOD

1. Use the capillary pipette provided with the Unopette to make a hole in the reservoir of the small disposable device that is also provided. Collect 20 µL of blood with the capillary pipette (Morton et al. 1993; Hoff 2000; Golde et al. 2005; Christensen et al. 2009) and place the blood in the device's reservoir, which contains 2 mL of lysis buffer. Mix the solution with a vortex mixer and keep it at room temperature until the count can be performed (within 6 h).
2. Assemble the hemacytometer by centering a coverslip onto it and placing the unit on a flat bench surface in preparation for counting the number of cells in a given sample.

3. Place a small amount (5–10 μL) of the suspension on a clean hemacytometer, sufficient to fill the area over the counting grid under the coverslip but not enough to "float" the coverslip or overrun the scored surface.

   ▸ *See Troubleshooting.*

4. Focus on the large central square of the grid (the square with the finest degree of etching); this large central square is divided into 25 smaller squares. Use the 40x objective lens of the microscope and 10x oculars to count the platelets in five of these smaller squares—the four corner squares as well as the centermost square. Keep the tally for each square separately. Include in each count platelets that touch the borders of only two sides of a given square (i.e., top and bottom or left and right). Do not include those touching the borders of the reciprocal two sides.

   We have found that this strategy provides a conveniently accurate way to count the platelets; however, consistency in the counting strategy is what is important—not the particular rules adopted. Typically, expect on average ≥20 platelets in each of the 25 boxes of the large central square of the hemacytometer.

   ▸ *See Troubleshooting.*

5. Determine the platelet count of the peripheral blood sample in question as follows:

   i. Average the five values from Step 4.

      ▸ *See Troubleshooting.*

   ii. Multiply the average platelet count by 25 (i.e., the total number of boxes in the large central square of the grid) to obtain the total number of platelets in the central square of the grid.

   iii. The large central square has an area of 1 mm^2 and a depth of 0.1 mm; therefore, multiply the value obtained in Step 5.ii by 10 to generate the number of platelets/mm^3 in the suspension that was loaded onto the hemacytometer.

   iv. Multiply the concentration of platelets in the suspension by the initial dilution factor used to generate the suspension (i.e., 1:100, if 20 μL of blood was diluted in 2 mL of lysis buffer in Step 1).

TROUBLESHOOTING

Problem (Step 3): Fluid "floats" the coverslip and/or there are bubbles between the hemacytometer and the coverslip.

Solution: If too much fluid has been applied, removing some of the volume with a towel or other absorbent tissue can skew the distribution of cells over the grid or disproportionately affect cell numbers. Likewise, bubbles will distort distribution; never use squares with bubbles for a count because the air displaces fluid volume that normally would contain platelets to be included in the count. If either of these occur, it is best to clean the hemacytometer and begin afresh.

Problem (Step 4): Platelets appear to be damaged.

Solution: Lysis from residual soaps or other contaminants can cause damaged samples. In addition, the use of an undiluted concentrated stock of suspension saline (e.g., 10x PBS) will result in shrunken platelets. Therefore, always begin with clean dry tubes, pipettes, and other materials, as well as properly balanced and diluted stock diluents.

Problem (Step 4): There are too few (<10) or too many (>50) platelets per square.

Solution: Given the estimates of normal platelet counts in mice (see the poster, Mouse Peripheral Blood Cells), the dilution steps should be planned to yield counts, on average, of 20–35 platelets per each of the five (four corner and center) squares indicated in Step 4. If counts are outside of this range, first verify that the dilutions were performed appropriately; if so, adjust the dilution steps to target the suggested range. The animal may be thrombocytopenic (too few platelets) or thrombocytotic (too many platelets), either of which will need to be documented.

Problem (Step 5.i): One or more of the counts obtained in Step 4 deviates by more than ~10% from the mean calculated in Step 5.i.

Solution: The numbers of platelets in the individual squares are counted and recorded independently to ensure an even distribution of the suspension in the hemacytometer. The individual values within the counted squares of the larger center square of the grid should not deviate by more than ~10% from the mean. If this occurs, disassemble, clean, and reassemble the hemacytometer before reloading the suspension for a repeat platelet count.

Protocol 4

Methodology to Determine a White Blood Cell Count and Viability Using a Hemacytometer

This protocol describes how to assess the number and viability of nucleated white cells in peripheral blood. A similar method can be used to enumerate nucleated white cells of spleen and bone marrow, with the final value expressed as number of cells per tissue as opposed to number of cells per unit volume. The value of the hemacytometer count method is that it also permits a direct visual assessment of the viability of the cell population being counted. This is especially necessary if the cells are being used for in vitro testing or are being isolated for engraftment into recipient animals, or if initiating tissue culture or clonogenic assays. The method of dye exclusion has been used to assess viability for decades and can be learned quite easily. Living cells retain a light-refracting quality owing to the integrity of the cell membrane. Dead cells lose this quality and appear dull, having taken up some of the dye. Thus, cells that exclude dye are living, and cells that take up dye are dead.

MATERIALS

CAUTION: See Appendix for proper handling of materials marked with <!>.

Reagents/Animals

HCl (0.1 N) <!> containing crystal violet <!>

Use a "pinch" (~10 mg) of crystal violet to prepare 250 mL of solution; it will have an aqua or blue-green color. Trypan blue <!> and other commercially available dye solutions work as well.

Mouse from which a blood sample is to be collected

Equipment

Bucket with ice

Compound light microscope with 16X objective lens and 10X oculars

Coverslips (glass)

Hemacytometer (Fig. 2)

Heparinized capillary tubes

If commercial preheparinized capillary tubes are not available, heparinize glass capillary tubes (which may be obtained from a variety of sources) by submersion in 1X PBS containing 20 units/mL heparin <!> for 1 h. Once treated, drain and air-dry the tubes before use. Prepared tubes may be stored indefinitely in capped containers en masse at room temperature.

Plastic tubes (3–5 mL)

Vortex mixer

METHOD

Some skill and experience are required to determine viable versus nonviable cells by dye exclusion under a microscope. Thus, we recommend controls with added dead cells as a means of acquiring this experience. A cell suspension standing for 2–3 hours at room temperature will have lost a considerable number of viable cells and may be used to learn the quality of dye uptake by nonviable cells.

1. Collect 50 µL of peripheral blood in a heparinized capillary tube (Morton et al. 1993; Hoff 2000; Golde et al. 2005; Christensen et al. 2009) and place it into a 3–5-mL disposable tube on ice that contains 1 mL of cold 0.1 N HCl with crystal violet. Mix thoroughly using a vortex mixer or by gently inverting the sealed tube.

   Crystal violet is used to determine cell viability by dye exclusion. In the presence of crystal violet, the peripheral blood suspension will turn dark brown, owing to release of hemoglobin following lysis of the RBCs.

2. Assemble the hemacytometer by centering a coverslip onto it and placing the unit on a flat bench surface in preparation for counting the number of cells in a given sample.

3. Place a small amount (~10 µL) of the suspension on a clean hemacytometer, sufficient to fill the area over the counting grid under the coverslip but not enough to "float" the coverslip or run over the scored surface.

   ▸ *See Troubleshooting.*

4. Use the 16X objective lens of the microscope with 10X oculars to count the WBCs in each of the four largest corner squares of the grid. Keep the tally for each square separately. In each count, include WBCs that touch the borders of only two sides of a given square (i.e., top and bottom or left and right). Count a cell if it touches the line, even if most of the cell is outside the square. Do not include those touching the borders of the reciprocal two sides.

   We have found that this strategy provides a conveniently accurate way to count the WBCs. Consistency in the counting strategy is what is important—not the particular rules adopted.

   ▸ *See Troubleshooting.*

5. Calculate the WBC count of the peripheral blood sample in question as follows:

   i. Average the four values from Step 4.

   ▸ *See Troubleshooting.*

ii. Each square has an area of 1 mm^2 and a depth of 0.1 mm; therefore, multiply the average WBC count by 10 to obtain the number of WBCs/mm^3 in the suspension that was loaded onto the hemacytometer.

iii. Multiply the concentration of WBCs in the suspension by the dilution factor used to generate the suspension (i.e., 1:20, if 50 μL of blood was diluted in 1 mL of 0.1 N HCl with crystal violet in Step 1) to determine the WBCs/mm^3 of peripheral blood.

TROUBLESHOOTING

Problem (Step 3): Fluid "floats" the coverslip and/or there are bubbles between the hemacytometer and the coverslip.

Solution: If too much fluid has been applied, removing some of the volume with a towel or other absorbent tissue can skew the distribution of cells over the grid, or disproportionately affect cell numbers. Likewise, bubbles will distort cell distribution; never use squares with bubbles for a count because the air displaces fluid volume that normally would contain cells to be included in the count. If this occurs, it is best to clean the hemacytometer and begin afresh.

Problem (Step 4): WBCs are dead.

Solution: Pay attention to the percent viable cells in the sample and make decisions according to the parameters of the experiment as to the degree of tolerance for nonviable cells. Normally, viability is expected to exceed 90%. Less than that may be a result of waiting too long between the preparation of the samples and counts, soap or solvent contamination of vials and tubes, or an inappropriate diluent. Therefore, always begin with clean, dry tubes, pipettes, and other materials, as well as properly balanced and diluted stock diluents.

Problem (Step 4): Too few (<10) or too many (>50) cells are included within each of the four corner squares to be counted.

Solution: The accuracy of the initial estimation of the sample's cell concentration is critical to the success of the hemacytometer count and depends on the dilution factor; i.e., if the initial dilution factor is too small, the number of cells applied to the hemacytometer will be too great and prevent an accurate count. In contrast, if the dilution factor is too large, too few cells will be present on the hemacytometer's grid, again preventing an accurate count. This is the reason why an approximation of the number of cells per unit volume of the initial suspension should be used to guide the dilutions for cell counting. Individual squares for a leukocyte count should contain between 10 and 50 cells for simple, accurate population estimates.

Problem (Step 5.i): One or more of the counts obtained in Step 4 deviates by more than ~10% from the mean calculated in Step 5.i.

Solution: The WBC numbers of the individual squares are counted and recorded independently to assure an even distribution of the suspension in the hemacytometer. The individual values within the four corner squares of the grid should not deviate by more than ~10% from the mean. If this occurs, repeat the count with a fresh preparation.

3 Peripheral Blood Films and Cytospin Preparations

TO PREPARE PERIPHERAL BLOOD FOR MICROSCOPIC EXAMINATION, two methods are generally available. The first is to spread a thin film of whole blood, drawn out from a single drop placed on a glass slide. This results in a spread that resembles a "feather" in a general sense, tapering off at the tip and limits. The spread will be a single-cell thickness with well-distributed leukocytes, platelets, and RBCs. All of the components are present because the spread is made using a drop taken directly from the animal. The other method involves using a cytocentrifuge. In this technique, the hypotonic treatment with distilled H_2O lyses all RBCs and platelets, and a diluted suspension of the remaining WBCs is centrifuged onto the surface of a slide. The most relevant difference between this method and a direct peripheral blood film is that the blood drop must be diluted substantially or mixed with anticoagulant as soon as possible to prevent the initiation of platelet activation and the clot formation cascade. The cytospin procedure is frequently used to examine aliquots taken from cell cultures and body fluid lavages.

An advantage of the cytospin method relative to the direct assessments of a peripheral blood film is that a defined number of WBCs is placed on the slide, all in the small area comprising the cytospun surface. That is, the density of WBCs in a cytospin will be much greater than that in a blood film, thus expediting examination; less time is spent examining the slide to accumulate the necessary number of differentiated cells for an accurate assessment. The primary disadvantages are that the number of nucleated cells proportional to the RBCs and platelets is lost, there may be a disproportionate loss of individual WBCs due to differential loss in the centrifugation steps, some structural integrity may be forfeited, and the process is labor intensive and time consuming.

The preparation of blood films, whether directly or by cytocentrifuge, is straightforward, but more art than science. A priori, as is true when using glass slides, a clean surface is required for best results. This chapter therefore begins with a description of a simple method to clean glass slides for use in the microscopic examination of blood, marrow, and other hematopoietic cells (Protocol 5). This is followed by a detailed description of methods to prepare a slide film of peripheral blood for microscopic examination (Protocol 6), procedures to fix, stain, and mount the blood film (Protocol 7), and steps to prepare slides of peripheral blood by cytocentrifuge (Protocol 8).

Protocol 5

Washing Slides in Preparation for Microscopic Examination of Peripheral Blood and Bone Marrow

The preparation of blood and bone marrow is straightforward, but a clean surface is required for best results. Although manufacturers describe the box contents as "precleaned," we have found that most glass slides have a thin residual film of unknown origin. To prepare truly good samples for microscopic examination, prewashing the slides is recommended. These steps are outlined here.

MATERIALS

CAUTION: See Appendix for proper handling of materials marked with <!>.

Reagents

Distilled H_2O
Ethanol (70% and 100%) <!>
Soapy water (warm/tepid, 37°C) (e.g., with Alconox <!>)
Tap water
Xylene <!>

Equipment

Glass slides (with frosted or colored ends)

Decide in advance whether to use slides with colored or frosted ends. It is frustrating to be ready for blood film or bone marrow preparation only to find that the slides have no provision for labeling!

Lint-free tissues (e.g., KimWipes)
Slide box for storing cleaned slides
Slide cassette with staining dishes or Coplin jar

METHOD

1. Load the slides into the cassette of a glass staining dish or other suitable container (slides may also be washed individually in a Coplin jar or other container). Make sure that the fluids can flow freely when the slides are immersed in the container.

2. Dip the cassette (or individual slides) three to five times (a few seconds each time) in the following sequence:

   xylene

   100% (absolute) ethanol

   70% ethanol

   warm/tepid (~37°C) mildly soapy water (e.g., with Alconox)

3. Drain the slides but do not air-dry. Then repeat the immersion in mild soapy water (last sequence step of Step 2).

4. Immerse the slides twice in tap water. Change the tap water after several sets of slides have been rinsed.

5. Immerse the slides twice in distilled H_2O. Change the distilled H_2O after several sets of slides have been rinsed.

6. Dip the cassette (or individual slides) three to five times (a few seconds each time) in the following sequence:

   70% ethanol

   100% (absolute) ethanol

7. Wipe the slides dry with lint-free tissues (e.g., KimWipes). Place the slides in a slide box and close the cover to keep them dust-free.

   Stored slides may be kept indefinitely at room temperature.

Protocol 6

Preparation of a Peripheral Blood Film

This protocol describes how to prepare a slide film of peripheral blood for microscopic examination. It is good practice to prepare two slides for each sample. This way, each slide may be used to spread the film of the other, and duplicate slides will be of value if something happens to one of them in the process.

MATERIALS

CAUTION: See Appendix for proper handling of materials marked with <!>.

Reagents/Animals

Chloroform (*optional*; see Steps 9–10) <!>
Compound light microscope with high-power oil immersion objectives
Immersion oil (*optional*; see Step 8)
Mounting medium (e.g., Permount)
  Dilute before use (two parts mounting medium to one part xylene <!>) (*optional*; see Steps 9–10).
Mouse from which a blood sample is to be collected
Silver nitrate powder (e.g., Kwik-Stop Styptic Powder) or gel (e.g., Kwik-Stop Styptic Gel Formula) <!>

Equipment

Coverslips (glass) (*optional*; see Steps 9–10)
Glass slides, washed and prepared as described in Protocol 5
  This procedure requires two slides per animal under study; each slide is used as a "draw" slide to prepare the film on the other slide.
Mouse restrainer
Pencil to label slides
  Do not use a pen or other marker material, because these may be dissolved by solvents during the staining process (i.e., all labeling will be lost).
Sealed box containing desiccant (e.g., Drierite) (*optional*; see Step 7)
Sterile scalpel, straight-edged razor blade, or sharp dissection scissors (per IACUC and/or veterinarian approval)
Swabs containing 70% isopropyl alcohol <!> or 70% ethyl alcohol <!>
  Alternatively, spray 2 x 2-inch squares of gauze with a spray bottle containing 70% isopropyl alcohol or 70% ethyl alcohol.

METHOD

1. Use a pencil to prelabel all slides on the frosted (textured) or colored surface. Label two slides per animal. Do not attempt to label the slides during bleeding; always label the slides in advance.
2. Set out the mice by caged groups so that they can be retrieved in any order.
3. Place a single drop of blood on the surface of the first labeled slide, near the frosted end (see Video 3, Blood Film).
4. Place the second labeled slide in contact with the first, so that the drop of blood is inside an acute angle between the two slides. If the drop of blood is scant, adjust the slide so that the angle between the two is as acute as possible. If the drop is particularly copious, increase the angle.

   ▸ *See Discussion.*

5. Working as quickly as possible, make the spread using a single, steady motion, never losing contact between the two slides (see Video 3, Blood Film).

   The blood will begin to clot, especially as it contacts the glass slide, so work as quickly as possible. If the blood is initially collected by vascular access using a syringe with an anticoagulant (e.g., EDTA, citrate, or heparin), timing is not as critical. However, there is a greater likelihood that the droplet will be larger, requiring more skill to make a well-distributed smear.

   ▸ *See Troubleshooting.*

6. Prepare a second smear by repeating Steps 3–5, using the edge of the first slide to spread a drop of blood on the second slide.

   It is not appropriate to use the same edge of a slide for more than a single film.

7. Air-dry the slides.

   The slides should be dried as quickly as possible to prevent hypertonic shrinkage of the cells, but the drying time is a function of ambient humidity, i.e., the higher the relative humidity, the longer the dry time for blood smears. Indeed, very humid conditions may necessitate the use of a sealed box containing desiccant (e.g., Drierite) to allow the smears to dry sufficiently for subsequent staining, fixation, and mounting as described in Protocol 7. It will be apparent that they are dry once the "wet" appearance of the spread disappears. Under normal circumstances, we recommend that staining be performed (as described in Protocol 7 and Video 4, Staining with Coplin Jar or Carriages) as soon as the smears are thoroughly dry (i.e., 15–30 min postpreparation). Alternatively, for a quick differential or visual exam, proceed to Step 8.

8. If necessary or desired, examine the stained blood films immediately after drying, without coverslipping, using high-power oil immersion objectives.

Although this step is quick and convenient, the blood films are vulnerable to damage and cannot be archived without mounting. Mounting medium and a coverslip provide protection for the film from environmental damage and decomposition.

▸ *See Troubleshooting.*

9. To coverslip the slide examined in Step 8, remove the immersion oil by dipping the slide in chloroform several times until the oil is obviously removed from the surface of the slide.

   Although immersion oil is very soluble in other less toxic solvents such as methanol, ethanol, or acetone, each of these quickly removes stain from the cells of the blood film, rendering the slide useless for cell differential analyses.

10. After removing the immersion oil, air-dry the slide and add a coverslip with mounting medium as described in Video 5, Coverslip.

    Slides coverslipped in this way will be well preserved for decades maintained at room temperatures in light-tight boxes.

TROUBLESHOOTING

Problem (Step 5): Blood clots before film preparation.

Solution: Consider the following:

- The interval between obtaining the drop of blood (Step 3) and spreading it on the slide (Step 5) may be too long. As skill develops, it will be possible to spread the film more rapidly.
- There may be clotting mechanism pathology in the mice under study. Although not common, it may occur and require collecting the blood initially in an anticoagulant such as EDTA or heparin to permit an acceptable spread.

Problem (Step 8): Blood films are too thick.

Solution: Blood sample size may have been too large. For large drops of blood, increase the angle between the top (spreading) slide and the slide containing the blood film as the spread is made. Alternatively, simply repeat the collection and take a smaller drop of blood.

Problem (Step 8): Blood films are too thin.

Solution: The blood sample size may have been too small. For small drops of blood, decrease the angle between the top (spreading) slide and the slide containing the blood film. In addition, begin the spread before the drop covers the entire distance of the contact between the two slides. Consider shortening the spreading stroke as well concentrating the smaller than average (i.e., limited) blood sample.

Problem (Step 8): Leukocytes are shriveled.

Solution: The blood film may have been air-dried for too long. Apply gentle heat and/or facilitate airflow over the slide to dry the film faster. Alternatively, reduce the volume of the blood sample used.

DISCUSSION

Visually inspecting a peripheral blood film provides important insights as to the quality of the blood film preparation. A suitably sized droplet and a steady motion with the top (spreading) slide are essential for good blood film preparation, creating a defined and extensive "feathered" area. Useful smears are shown in Figure 3, A and B, even though one of the smears (Fig. 3B) was initiated with a larger sample of blood. By comparison, other smears in Figure 3 were made with samples that were either too large (Fig. 3G) or too slight (Fig. 3H). This will make examination of the samples difficult due to cell crowding or the lack of available cells for examination, respectively. The blood smear in Figure 3C results when a slight amount of the blood droplet is ahead of the leading edge of the moving (top) slide. This occurs when the initial placement of the drawn slide is incorrect. Instead of being drawn back to the drop, the slide is placed at the drop's edge, inadvertently intersecting it such that a fraction of the blood is in front of the slide. The smear in Figure 3D results when the smear movement is halting and interrupted. Similarly, the smear in Figure 3E results from an irregularly paced motion with the upper slide. The smear in Figure 3F may simply be from too scant a drop, one in which there is an excess of interstitial fluid, or a smear from an anemic or otherwise sick animal.

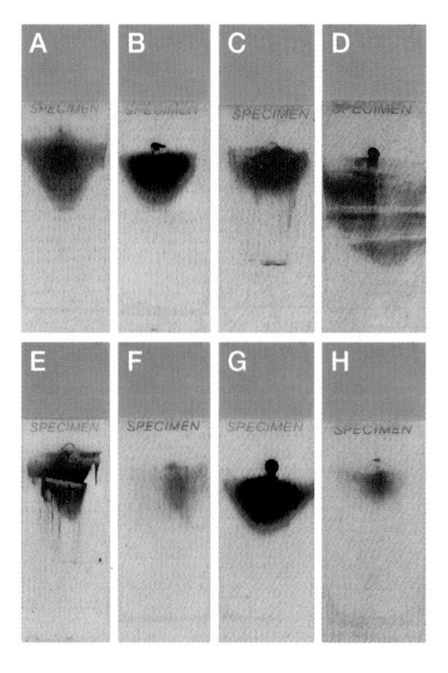

FIGURE 3. Successful preparations (*A*,*B*) and representatives of various unsatisfactory results (*C*–*H*) of blood film preparation. *C* occurs as a consequence of the top (spreading) slide "catching" a portion of the blood droplet ahead of the slide's leading edge. *D* results when there is a halting motion, interrupting the smooth extension of the droplet. *E* results from an irregular shift during the spreading stroke of the top slide. *F* occurs if the initial droplet is too slight, if the blood is diluted with interstitial fluid or serum, or if the animal is anemic. *G* is an example of starting with too large of a droplet of blood. *H* is an example of starting with too small of a droplet of blood.

Protocol 7

Fixing, Staining, and Mounting a Peripheral Blood Film

In this protocol, blood films are stained with a commercially available blood-staining kit. Despite some differences in details, all of the kits consist of a typical Romanowsky-dye combination of methylene blue and eosin preceded by methanol fixation. These are the basis for the Giemsa, Wright, and Wright–Geimsa stains and produce satisfactory results. Methylene blue (which is basic) stains acidic molecules (pH <7) blue/purple. Thus, the nuclei of cells, rich in nucleic acids, stain dark blue/purple. In contrast, eosin (which is acidic) stains basic molecules (pH >7) magenta/orange/pink. This includes, for example, the cytoplasmic granules in eosinophils that contain lysine- and arginine-rich proteins such as major basic protein-1 (MBP-1). We recommend an "old-school" mounting approach, using xylene-miscible media that may be diluted to some degree before use (e.g., two parts mounting media with one part xylene). This solvent–media mixture has virtually no interaction with the stains typically used to examine WBCs, and the process has remained virtually unchanged for 100 years, facilitating direct comparisons of samples with historical records in the literature.

MATERIALS

CAUTION: See Appendix for proper handling of materials marked with <!>.

Reagents

Blood-staining kit

A number of manufacturers make "3-bottle" blood-staining kits (e.g., Diff-Quik from Baxter Scientific Products or LeukoStat from Fisher Scientific). If a more intense stain is desired, use a Dominici stain (McClung 1961) or other special stains for specific cell types or subcellular components (Lillie 1977).

Methanol (100%) <!>

Alternate fixatives (e.g., Zenker-formol <!>) may be used if their benefits are needed for the preparations in question.

Mounting medium (e.g., Permount), diluted before use (two parts mounting medium to one part xylene <!>)

A variety of options exist for mounting media, including those with aqueous-based solvents; Web searches will identify a variety of alternative materials.

Xylene <!>

Equipment

Compound light microscope for assessing the slides after mounting

Coverslips (glass)

If only the feather edge is required, 22 x 22-mm-square coverslips may be sufficient. However, 24 x 50-mm coverslips are easier to handle and allow for examination of the entire spread if this is necessary or desired. Most objectives are designed for the standard coverslip thickness of 1.7 mm (1.5 oz.); other thicknesses can affect image quality. Thus, in our opinion, although 1-oz. coverslips are adequate, the thicker 1.5-oz. coverslips are slightly better for photography.

Fine forceps

Shaker for agitating slides

Slide box for storing slides

Slide rack or Coplin jar

Slides of peripheral blood films

Prepare as described in Protocol 6.

Warming table (*optional*; see Step 8)

Wipes, soaked in solvent

METHOD

It is always best to stain the blood films as soon after preparation as possible, when the films are thoroughly dry (i.e., 15–30 min postpreparation). The presence of even small amounts of moisture will distort the appearance of cells and limit the archivability of the slides. Most slides may be stained without loss of definition if stained within ~3 h. If a longer delay is necessary, slides may be kept for ~1 d at 4°C or fixed as described in Steps 1 and 2 below with 100% methanol for 30–40 min and then stained at a later time (up to 7–10 d later if kept at 4°C). However, make sure to stain and coverslip the slides as soon as possible after preparation.

Fixing

1. Place the slides in a rack or Coplin jar, frosted end up, all facing the same direction. Make sure that there is sufficient space between slides to permit fluid flow at the blood film surface.

   If the carriage and/or jar has slots that permit double-slide thickness, slides may be placed "back to back" (film surface out), essentially doubling the capacity. In this event, agitation of the fluid during staining by movement of the carriage and/or rocking of the jar is critical to expose the film surfaces to the fixative and staining solutions. Take extra caution when removing slides placed back to back to avoid scraping the blood film surface.

2. Submerge the rack in 100% methanol (or fill the jar with 100% methanol) and incubate for 30–40 min at room temperature with occasional agitation to ensure that the film is bathed by the methanol.

3. After 30–40 minutes, remove the slides and allow them to air-dry.

   The slides may be kept in a light-tight slide box for 7–10 days if stored at 4°C.

Staining

4. Stain according to the manufacturer's instructions associated with the blood-staining kit selected for use (see Video 4, Staining with Coplin Jar or Carriages).

   Stained slides may be kept uncovered, although it is best to coverslip them within ~1 d as described in Steps 5–8.

Mounting

The application of coverslips must be performed swiftly. Usually, ~4–6 slides may be processed at one time; more than this amount of slides will invariably affect the quality of the final slides.

5. Dip completely dry slides in xylene in a chemical fume hood and place the blood film side up on the bench surface.
6. Apply a small drop (~50 μL) of mounting medium to a 22 x 22-mm coverslip (proportionally more medium is used with larger coverslips) and immediately "drop" the coverslip onto the surface of a freshly xylene-dipped slide (see Video 5, Coverslip).

   When mounting medium is exposed to the air, the solvent immediately begins to evaporate, increasing its viscosity, eventually to an unworkable degree. Because the task must be performed in a chemical fume hood as a consequence of using hydrocarbon-based solvents, evaporation is even more rapid.

7. Grossly examine the preparations and remove any trapped air bubbles by gently depressing the coverslip with a fine forceps, shepherding bubbles under the coverslip to the outer edges of the slide. Remove excess medium by wiping it off with a solvent-soaked wipe.
8. Once coverslipped, allow the slides to dry and the medium to harden before visualizing the films under a microscope.

   It is not necessary to use a warming table, although some prefer to do this. The liability is that the table may get too warm and thereby alter the fine detail of the cells in the preparation. Slides coverslipped in this way and maintained at room temperatures in light-tight boxes will be well preserved for decades.

   ▸ *See Troubleshooting.*

TROUBLESHOOTING

Problem (Step 8): Many cells are lysed.

Solution: If cell lysis is lineage specific, it may be pathognomonic of a unique disease. If the effects are not lineage specific, consider systemic technical reasons such as residual soap or fingerprints on the slides.

Problem (Step 8): Staining of the cells is too light.

Solution: There are several possible causes: (1) Bottles of stains may have passed their expiration date. (2) Working stocks of stains may be too old and are no longer effective. (3) Spillover from one stain to another may have diluted individual stocks. (4) Slides may have been transferred too quickly from one stain to another. (5) Slides may not have been agitated sufficiently to allow adequate bathing of the films with the stains. This requires either sufficient "vigor" or agitation for adequate time to allow replacement of the previous solution by the current one at the surface of the slide. Regardless of the cause, wash the slides extensively in 70% ethanol to remove any existing stain, rinse them in deionized H_2O, and then repeat the staining procedure with fresh stocks of stains. Consider a final wash of slides in tap water as opposed to deionized H_2O to increase staining intensity.

Problem (Step 8): Staining of the cells is too dark.

Solution: There are several possible causes: (1) The staining stock solutions may have been left open and evaporation has concentrated the stocks. Fresh stock stains may be needed. (2) The slides may have been kept too long in the individual staining solutions. Decrease the duration in each stain. (3) The slides may not have been sufficiently rinsed in deionized H_2O. If this is the case, increase the number and/or duration of rinses in deionized H_2O.

Problem (Step 8): Small grains appear on the slides.

Solution: This effect may be from a precipitate that sometimes forms in stains. If this occurs, filter the individual staining stock solutions through filter paper before use.

Problem (Step 8): Smears appear blotchy.

Solution: Blotches are often indicative of old staining solutions. In other cases, the fixation step may have been insufficient; if the old fixative was left out, airborne moisture may have been absorbed by methanol. If this is the case, prepare new fixative and/or staining solutions.

Problem (Step 8): Coverslip does not rest evenly on the slide.

Solution: This may be caused by the presence of air bubbles or simply too much mounting medium. It may be remedied by expressing the bubbles as described in Step 7 or by removing excess mounting medium by removing the coverslip with a xylene soak and reapplying a reduced amount of mounting medium.

Problem (Step 8): Cells appear refractory under the microscope or in photographs.

Solution: The sample may not have sufficiently dried before dipping in xylene and coverslipping. If this occurs, remove the coverslip, and following a quick dip in xylene, rinse the slide extensively with 100% ethanol (at room temperature) until the surface of the slide no longer indicates the presence of water residue, rinse the slide twice in fresh xylene, and reapply the coverslip.

Problem (Step 8): Stained cells on the slides fade with time.

Solution: Excessive and/or intense light will bleach the stained cells on a slide. Store archived slides in the dark (a slide box or other container works well).

Protocol 8

Preparation of a Peripheral Blood Suspension for Cytospin Analysis

Specific steps for the preparation of a peripheral blood sample for cytocentrifugation are outlined here. These steps allow several slides, each with ~30,000 cells/slide, to be prepared from a typical blood sample.

MATERIALS

CAUTION: See Appendix for proper handling of materials marked with <!>.

Reagents/Animals

Distilled H_2O, prechilled on ice
Dulbecco's phosphate-buffered saline (PBS) (10×)
Dulbecco's PBS (1×) containing 2% fetal calf serum and 20 units/mL heparin <!>, prechilled on ice
Mouse from which a blood sample is to be collected
Saline solution containing 5%–10% fetal calf serum for wetting slides before cytocentrifugation (see Video 6, Cytocentrifuge Procedures)
Trypan blue solution (0.4%) <!>

Equipment

Bucket with ice
Clinical centrifuge or equivalent at 4°C
Cytocentrifuge and cytospin slide apparatus (see Video 6, Cytocentrifuge Procedures)
Glass capillary collection tube
Hemacytometer
Microcentrifuge tubes (1.5 mL)
Transfer pipettes

METHOD

1. Collect 20–30 µL of peripheral blood from the tail vasculature using a glass capillary collection tube (Morton et al. 1993; Hoff 2000; Golde et al. 2005; Christensen et al. 2009).

2. Quickly transfer this peripheral blood sample to a 1.5-mL microcentrifuge tube preloaded with 1 mL of ice-cold 1x PBS with 2% fetal calf serum and 20 units/mL heparin. Keep the blood on ice until all samples have been collected.
3. Centrifuge the sample at low speed (2000 rpm, ~1000*g*) for 5 min at 4°C to pellet the RBCs and WBCs.
4. Using a transfer pipette, aspirate the supernatant, being careful not to disturb the underlying cell pellet.
5. Gently tap each tube to disperse the cell pellets with the liquid that remains in the tube. Resuspend each dispersed cell pellet with 1 mL of ice-cold distilled H_2O by inverting the tube three to five times and then incubate the cells for 1 min on ice.

   The hypotonic nature of this treatment results in the lysis of all of the RBCs in the sample, with only nominal effects on the WBCs.
6. Add 100 µL (i.e., 1/10 the volume) of 10x PBS to stop RBC lysis.
7. Recover the WBCs by centrifuging at low speed (2000 rpm, ~1000*g*) for 5 min at 4°C.
8. Aspirate the supernatant, being careful not to disturb the underlying cell pellet.
9. If any residual RBCs remain (i.e., the cell pellet displays a red/pink hue), repeat the RBC lysis procedure (Steps 5–8).
10. Resuspend the final WBC cell pellet in 100 µL of ice-cold 1x PBS containing 2% fetal calf serum and 20 units/mL heparin.
11. Use a hemacytometer to count the WBCs from a 10-µL aliquot that has been diluted with an equal volume (i.e., 10 µL) of a 0.4% trypan blue solution (see Protocol 4) and calculate the concentration of WBCs in the suspension.
12. Dilute the WBC suspension with 1x PBS containing 2% fetal calf serum and 20 units/mL heparin so that the concentration is ~240 cells/µL (or ~30,000 WBCs in each 100–150-µL aliquot used to prepare a slide).
13. Perform cytocentrifugation (i.e., cytospin) as shown in Video 6, Cytocentrifuge Procedures, with two slides per blood sample. Adhere to the instrument's operation manual.

    Put effort into defining the technical parameters of the cytospin device. Variables such as the speed and duration of the spin, as well as the initial volume of cell suspension, should be maximized for the different cell populations being studied. This way, one can confidently report reproducible results from this technique.
14. Fix, stain, mount, and visualize the cells as described for a peripheral blood film (see Protocol 7).

    ▸ *See Troubleshooting.*

TROUBLESHOOTING

Problem (Step 14): Too few or too many leukocytes are present on the cytospin preparation. If there are too few cells, it may be difficult to find the cells, and the number of lysed or fragmented cells increases significantly. If there are too many cells, the overlapping and crowding may change the morphology of the cells as well as their staining properties.

Solution: Review the calculations and/or cell counts used to prepare the cell suspension for cytocentrifugation. The number of cells suggested for a cytospin varies greatly with "reliable" sources, which suggest as little as a few thousand to as many as 100,000 cells. In our experience, the window for the appropriate number of cells is much smaller, with a focus on 30,000 cells.

4 Cell Differential Assessments of Peripheral Blood Films

ANY PERIPHERAL BLOOD NUCLEATED WHITE CELL differential analysis should be initiated with a cursory examination of the blood film from a global perspective. A remarkable amount of information may be gained from an initial overview of a blood film at lower powers of magnification (e.g., 10× objective with 10× oculars). The blood film should initially be screened for abnormal distributions, streaks, clumping, and other irregularities (Fig. 4). Streaks of purple material (DNA from disrupted leukocytes) may be expected in small measure. Similarly, indistinguishable denuded cells (nuclei without cytoplasm) may be encountered. However, if denuded cells comprise ≥5% of the cells in the blood film, something is likely to be wrong with the procedure or it may be pathognomic of a condition in the animal relating to cellular fragility. This is especially true if the observation is lineage specific. Likewise, the incidence of histiocytes, endothelial cells, or other tissue/vascular cells may be reflective of an "aggressive" preparation of the blood drop for the smear and may be a reason to reject the slide.

The cell differential is best performed within the feather edge of the blood film. The boundaries of the tip and side margins of the smear are not appropriate because the cellular distributions in these regions are often skewed. After beginning, the cell count should be continuous and across contiguous fields of view until at least 100, and preferably ≥200, cells are counted. This is done by scanning across the film, from side to side within the film margins, and then adjusting the slide to an adjacent (but not overlapping) field so as to avoid replicate counting of individual cells (Fig. 5).

The accompanying poster (Mouse Peripheral Blood Cells) and Figures 6–11 present a variety of high-power light microscopic views that are representative of each of the formed elements of the peripheral blood. We suggest using these as a quick reference guide to facilitate the cell differential analysis of nucleated white cells. The development of true expertise to perform cell differential assessments is based on repeated examination of blood films. We particularly suggest that investigators with little experience seek advice from hematology and/or comparative medicine colleagues to gain the necessary experience to distinguish among these cells

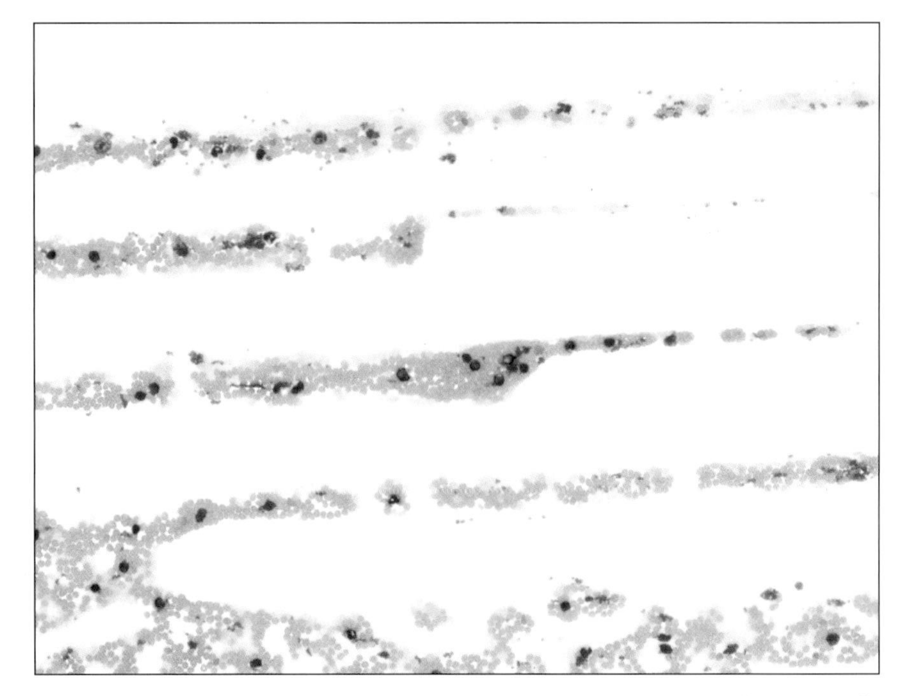

FIGURE 4. "Streaked" area of a film showing "fingers" of blood that will occasionally extend from the body of the spread blood. Although common and a potential source of many WBCs, it is not advisable nor good practice to use these areas for a differential assessment of the film because there may be a disproportionate retention of cells of some lineages.

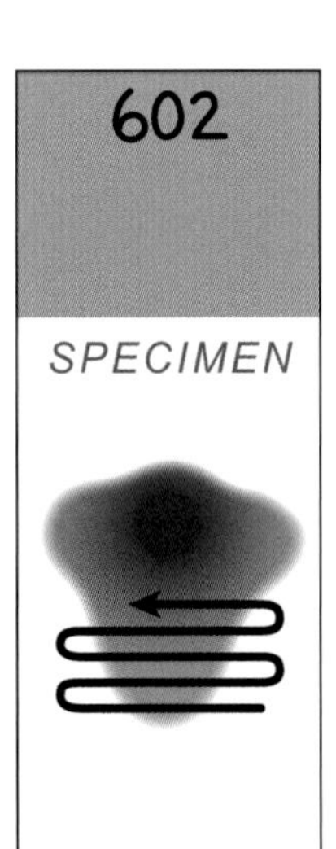

FIGURE 5. Schematic representation of the method needed to scan a blood film for differential cell analyses. Once the feather edge is located, the cell differential assessment is accomplished by a steady progression across the smear; i.e., the investigator should move from one contiguous field to the next, making sure to classify each cell as it is encountered. On the return "sweep" of the slide, it is necessary to have moved the field of the objective so as not to overlap previously examined microscopic fields.

before systematic errors creep into cell differentials (i.e., practice without training and guidance does not make perfect—it makes permanent).

The nucleated cells (leukocytes) and other formed elements of peripheral blood normally exist within a limited range in healthy animals. It is thus possible to gain an initial impression of whether the blood values are relatively normal by comparing the cell counts to one another using the 40X objective and 10X oculars.

Technical Tip: Generally, the red blood cell (RBC) count range for a mouse is 7×10^6 to 13×10^6/mm^3, whereas the platelet range is 3×10^5 to 10×10^5/mm^3. Thus, one can expect to find about one platelet for every 10 RBCs. However, platelets are only randomly disbursed if the blood sample has been treated with an anticoagulant. Otherwise, they will occur in clusters and clumps, making estimates of their numbers impossible. The cell count range of the remaining component of peripheral blood, the white blood cells (WBCs), is 5×10^3 to 12×10^3/mm^3 in a healthy mouse maintained in most pathogen-free environments. Thus, there should be about one nucleated WBC for every 10^2 platelets and every 10^3 RBCs. Although these values are based on ranges, the approximations can be valuable for recognizing one order of magnitude alteration in the ratios of the different peripheral blood components.

The formed elements of blood are mainly *erythrocytes* (i.e., RBCs, the classical hemoglobin-containing biconcave discs), which are all shaped approximately the same (isomorphic) (Fig. 6). Anisocytosis (i.e., red cells of varying shapes and sizes) is a prognostic indicator of pathology. RBCs that have been recently released from erythropoietic tissues are slightly larger and retain a bluish-gray tinge, reflective of residual RNA associated with hemoglobin synthesis; they are called *reticulocytes.* As reticulocytes mature, the slight basophilia owing to the presence of this RNA is lost and the cells acquire the normal red color of circulating RBCs.

Technical Tip: The identification and quantification of reticulocytes in circulation provide a relatively quick and informative assessment of the subject mouse, including its age. In an otherwise healthy adult animal, reticulocytes comprise ~3% of total RBCs. However, this number is larger in younger animals (e.g., 7%–10% at weaning), and increased numbers of circulating reticulocytes (e.g., >10%) are a prognostic indicator of trauma (e.g., a sudden and significant loss of blood), diseases affecting erythopoiesis (e.g., renal tumors that release erythropoietin), and conditions resulting in hypoxia.

Clusters of *thrombocytes* or platelets (small heterogeneously staining and irregularly shaped objects appearing more as cellular fragments than as intact cells) (Fig. 6), as well as cells of the various lineages of circulating nucleated WBCs are also present. WBCs generally fall into two distinct subtypes: (1) polymorphonuclear leukocytes of granulocytic lineages, including *neutrophils* (Fig. 7), *eosinophils* (Fig. 8), and *basophils* (rare) (Fig. 9) and (2) lymphomononuclear cells such as *lymphocytes* (Fig. 10) and *monocytes* (Fig. 11). The WBCs in peripheral blood should all be mature. WBC numbers are scant in cases of leukopenia and abundant during leukemoid reactions and in some leukemias.

Erythrocytes

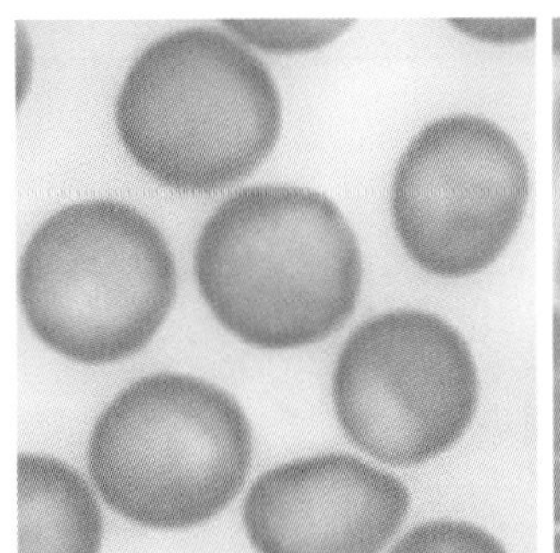
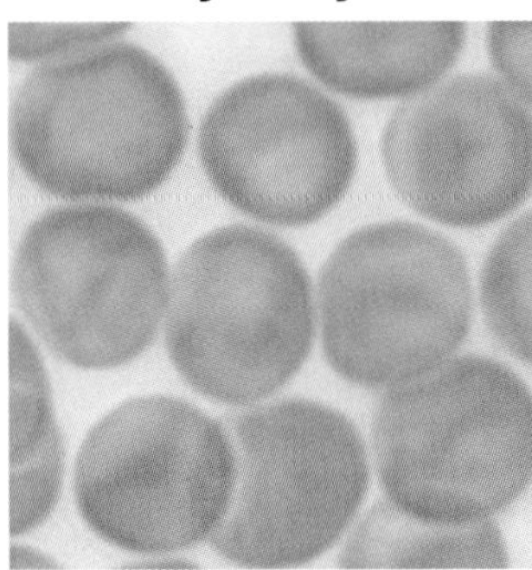
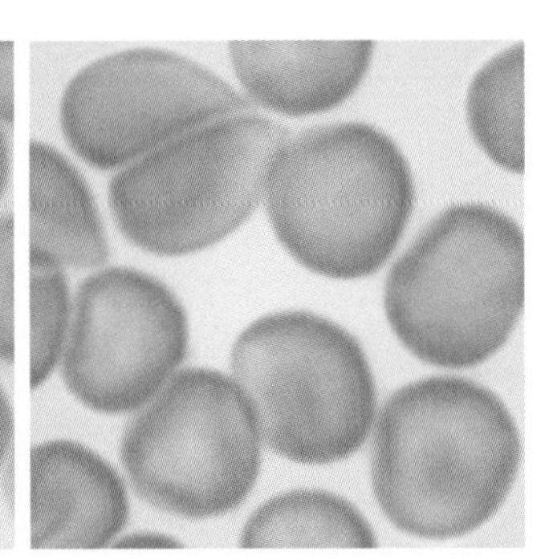

Thrombocytes

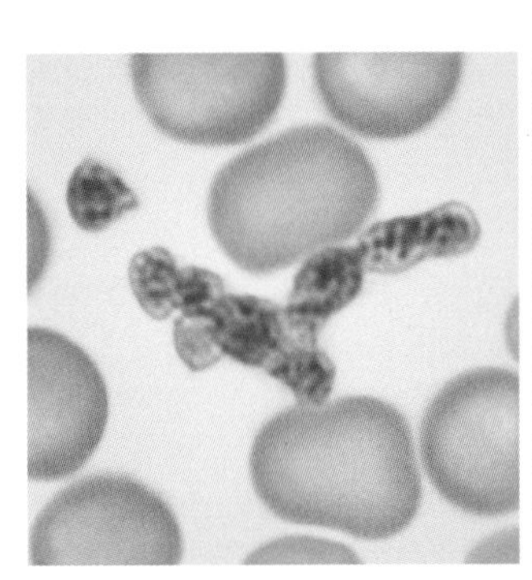
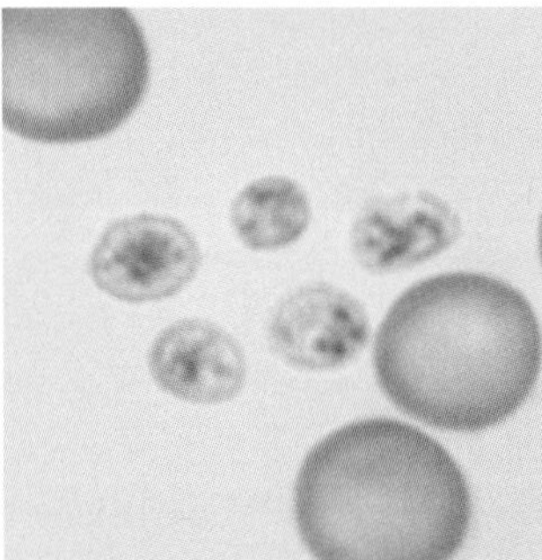
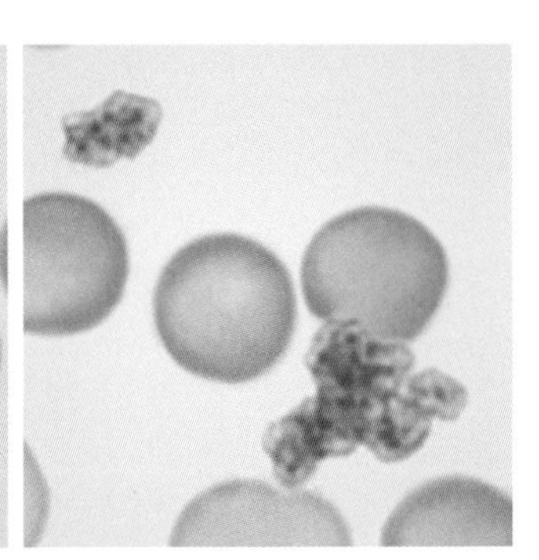

FIGURE 6. Variation observed in peripheral blood erythrocytes and thrombocytes. Erythrocytes (i.e, RBCs) and thrombocytes (i.e., platelets) comprise the majority of blood cells, at 0.7×10^7 to 1.3×10^7/mm^3 and 3.0×10^5 to 10.0×10^5/mm^3, respectively. In contrast, nucleated WBCs (i.e., leukocytes) comprise the smallest fraction at only 5.0×10^3 to 12.0×10^3/mm^3. Erythrocytes display little variation from their 6–8-μm diameter, concave disc-like appearance. The only variation of note is the occasional cell with a basophilic (i.e., blue-gray) hue, representing a recently formed cell from the marrow (i.e., a reticulocyte), whose residual cytoplasmic globin RNA results in the basophilic appearance. Thrombocytes are anucleate cytoplasmic fragments of marrow-resident megakaryocytes that are much smaller in appearance (1–2 μm) relative to either erythrocytes or leukocytes. This small size prevents visualization of significant variation among thrombocytes (regardless of the magnification), other than noting the appearance of single cells or clumps/groups of cells.

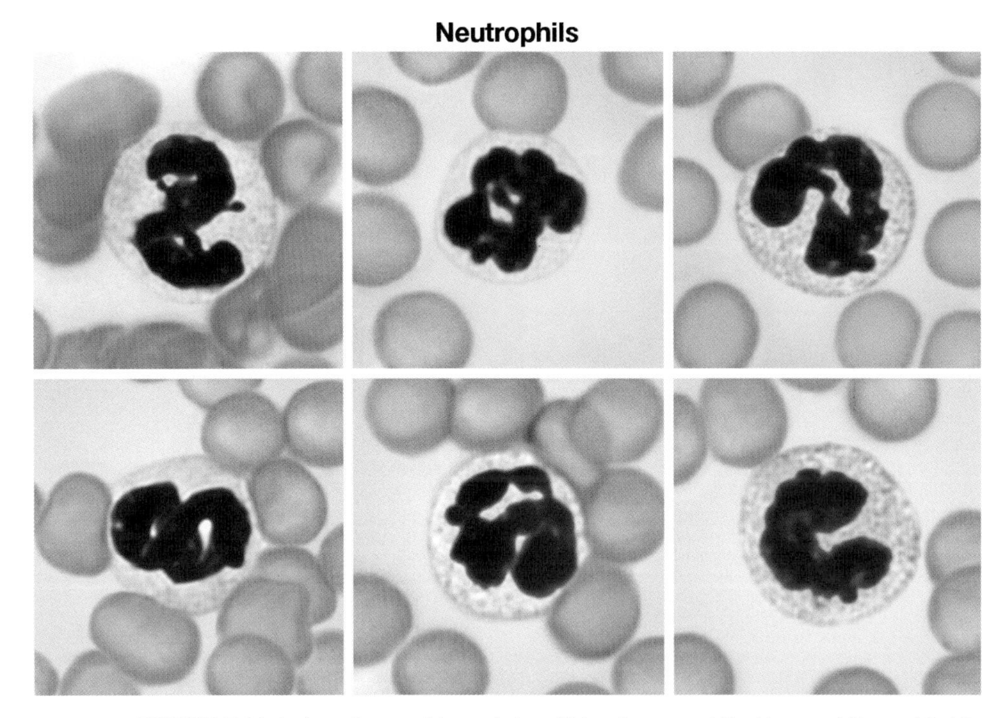

FIGURE 7. Variation observed in peripheral blood neutrophils. Neutrophils are 12–15-µm nucleated WBCs that generally comprise 20%–25% of the peripheral blood leukocytes in a mouse. Only mature cells are found in the blood of healthy animals, and they display variation in both nuclear and cytoplasmic morphology. In particular, the nuclei of neutrophils are indented or multilobed with clumped chromatin regions, extending from highly heterochromatic "strand-and-lobe" appearances to less-condensed chromatic regions as observed in "C-type" nuclei. Cytoplasmic variation observed among these cells includes a staining appearance that is lilac pink to clear that often (but not always) contains plentiful fine pale or neutral-colored granules.

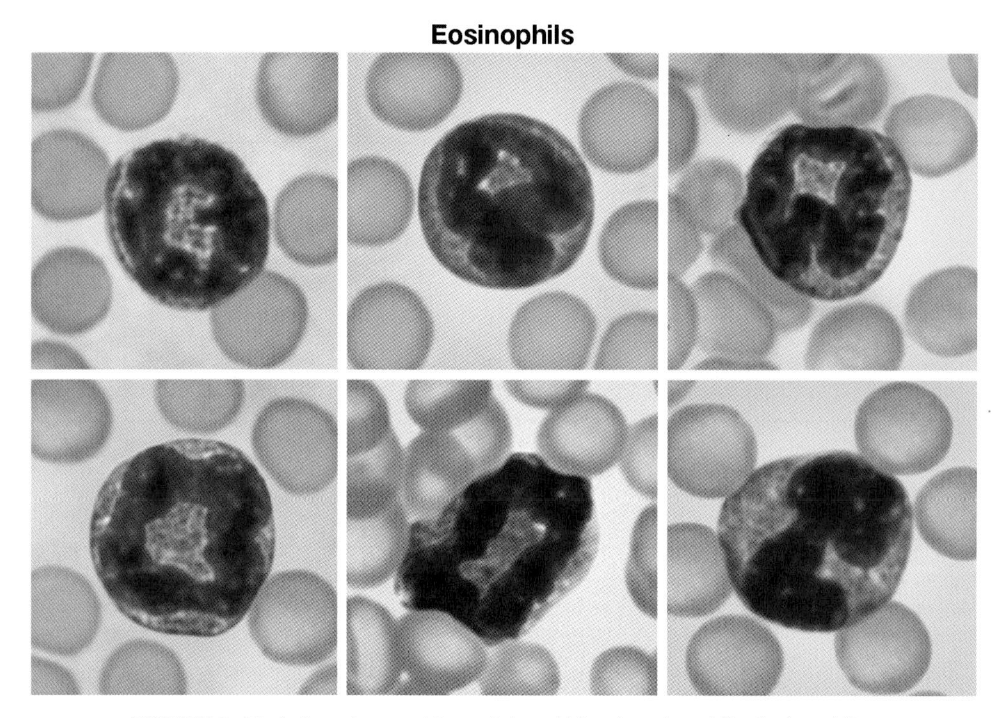

FIGURE 8. Variation observed in peripheral blood eosinophils. Eosinophils are 12–15-μm nucleated WBCs that generally comprise 0%–3% of the peripheral blood leukocytes in a mouse. Only mature cells are found in the blood of healthy animals, displaying variation in both nuclear and cytoplasmic morphology. In particular, the nuclei of mouse eosinophils invariably display a ring and/or doughnut shape that in many cases twists to form a "figure eight." Collectively, cells with these nuclear configurations comprise >95% of all blood eosinophils, with the remaining eosinophils displaying nuclei that are either segmented or multilobed or that have a ring nucleus that is simply viewed edge-on. In general, chromatin regions of the nuclei display significantly less condensation relative to neutrophils. The cytoplasmic regions of eosinophils display staining appearances whose orange/red/pink intensities may vary based on the density of the uniformly sized secondary granules in a given cell.

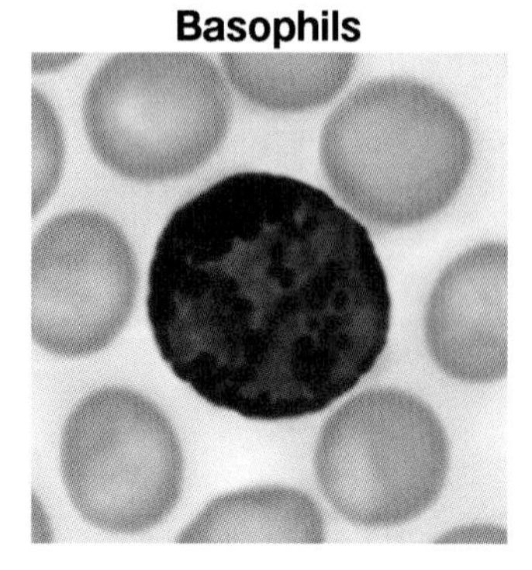

FIGURE 9. Variation observed in peripheral blood basophils. Basophils are 12–15-μm cells that are extraordinarily rare in the peripheral blood of mice (<0.1% of all leukocytes). The nuclei of these granulocytes are segmented and are often obscured or overlaid by dense, coarse granules that have a very dark blue to black staining appearance. Cytoplasmic variation in these cells includes the presence of granules with different sizes and shapes. Moreover, it is not uncommon to observe cells with only a small to moderate number of granules.

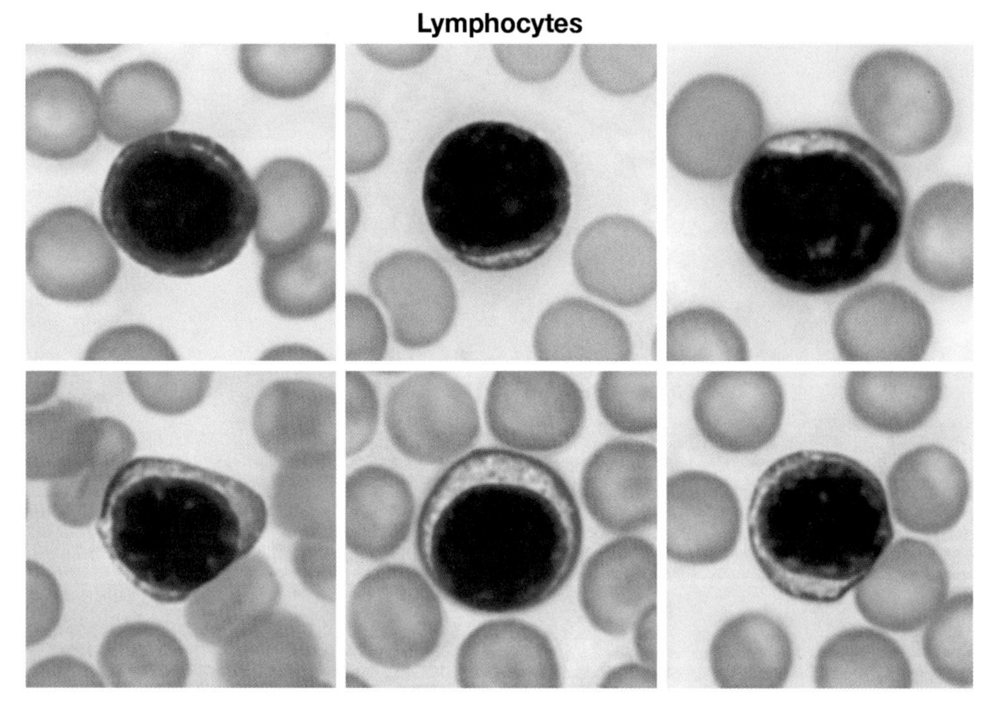

FIGURE 10. Variation observed in peripheral blood lymphocytes. Lymphocyte size varies but is nonetheless generally smaller than granulocytes, displaying a size range of 8–15 µm. However, these leukocytes are by far the most prevalent nucleated WBCs in the mouse, comprising 70%–75% of the peripheral blood. Only mature cells are found in the blood of healthy animals, and they display variation in both nuclear and cytoplasmic morphology. In particular, the nuclei of mouse lymphocytes invariably are small, compact, and round or oval in shape with a coarse appearance and few extensively heterochromatic areas; these cells are "mononuclear" cells and as such their nuclei never display the filamentous structures often associated with granulocytes. Furthermore, the nuclei of lymphocytes are also almost always asymmetrically oriented in the cytoplasm, usually off-set within the cell, and "resting" on a portion of the inner surface of the plasma membrane. The staining hue of the cytoplasmic regions of lymphocytes varies a little but is generally basophilic (i.e., vibrant blue) with little to no granulation. The very large nucleus-to-cytoplasmic ratio of these mononuclear leukocytes provides yet another defining characteristic to distinguish these cells from peripheral blood monocytes.

Monocytes

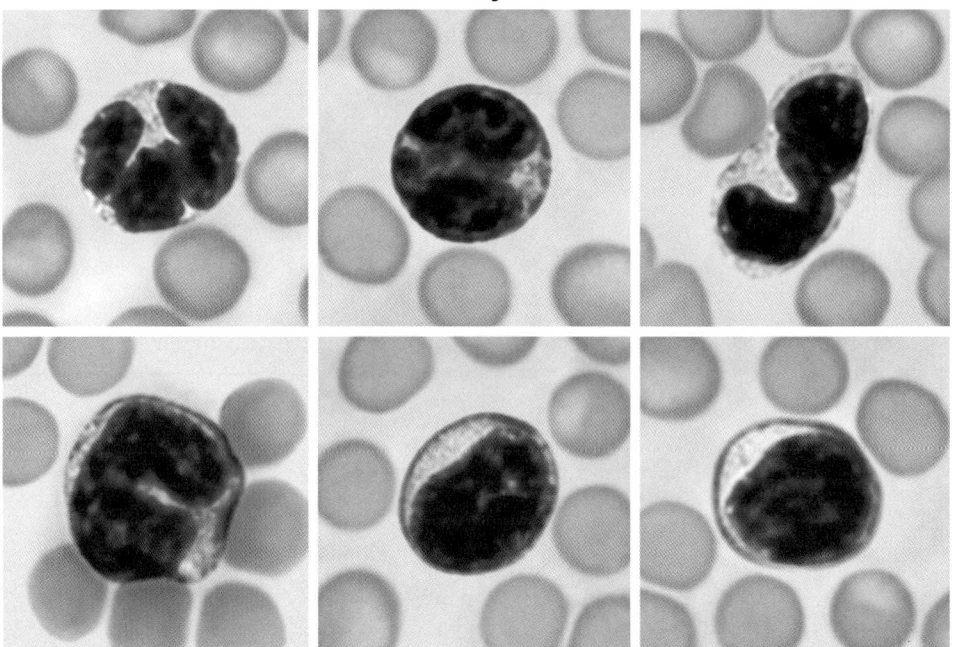

FIGURE 11. Variation observed in peripheral blood monocytes. In general, monocytes are slightly smaller than granulocytes but at the same time they are slightly larger than lymphocytes, displaying sizes between 10 and 15 μm. These leukocytes comprise a somewhat small nucleated WBC population in the mouse (2%–6% of the peripheral blood). Only mature cells are found in the blood of healthy animals, and they display variation in both nuclear and cytoplasmic morphology. In particular, the nuclei of mouse monocytes are indented or lobular in appearance with "lacy" heterchromatic areas; these cells are "mononuclear" cells and as such their nuclei never display the filamentous structures often associated with granulocytes. However, unlike lymphocytes, the nuclei of monocytes are not necessarily round—a characteristic that, when present, allows these cells to be easily distinguished from the more prevalent mononuclear lymphocytes. Furthermore, the nuclei of monocytes are also invariably oriented toward the center of the cell with at least a scant strand of cytoplasm separating the nuclei from the plasma membrane. The staining hue of the cytoplasmic regions of monocytes is somewhat muted relative to lymphocytes; thus, monocytes are often less basophilic (i.e., blue) with little to no granulation. Another defining characteristic of monocytes is a very large nucleus-to-cytoplasmic ratio that is nonetheless significantly smaller than the ratio observed in lymphocytes, providing yet another defining characteristic to distinguish these mononuclear leukocytes.

Did you know? Although the WBC counts of otherwise healthy mice and humans are nearly equivalent (5×10^3 to $12 \times 10^3/mm^3$ of blood), a hallmark species-specific distinction between humans and mice lies in the relative numbers of neutrophils versus lymphocytes. Neutrophils and lymphocytes are codominant WBCs of humans, each comprising >40% of all peripheral WBCs. In contrast, lymphocytes are singularly the dominant WBCs of mice, often comprising >70% of all peripheral WBCs; neutrophils only comprise ~20% of mouse peripheral WBCs.

The unique morphologies characteristic of mature nucleated cells in peripheral blood are used to distinguish and categorize specific leukocyte lineages in the blood. Specific and detailed descriptions of WBC morphologies gained from our experiences are provided in the descriptions of the cellularity associated with hematopoietic compartments (see Chapter 6). However, it is critical to recognize that only mature forms of each WBC should be present in the peripheral blood of an otherwise healthy mouse. Mitotic WBCs are abnormal, as are cells of considerable lesser maturity. Immature WBCs are normally found only in hematopoietically active sites such as the marrow or spleen of the mouse. Evidence of immature and/or blast forms of leukocytes in the blood indicates significant physiological stress and/or disease; effects on unique leukocytic lineages may provide valuable insights into the origin(s) of this stress and/or disease.

Did you know? Circulating WBCs can be described as showing a "left shift" of the lineage affected, owing to the tradition of drawing the differentiation pathway from primitive cells to mature cells in a left-to-right direction. Thus, a shift of cells to a lesser degree of maturity would be "to the left."

5 Preparation of Bone Marrow for Microscopic Examination

PERHAPS BECAUSE OF ITS COMPLEXITY and relative inaccessibility, the preparation and subsequent microscopic examination of hematopoietic tissues, especially bone marrow, are not routinely performed in research labs using mouse models. Thus, an essential component of all responses that involves blood cells is lost to analysis. In an effort to correct this, we endeavor here to outline simple techniques to enable the preparation of representative samples of marrow cellularity for microscopic examination.

One procedure, brush smears (Protocol 10), directly samples from the marrow milieu. The other procedure, cytospin preparations (Protocol 11), is perhaps more familiar, but it requires the preparation of a suspension and a centrifugation step, either or both of which may involve selective loss of cells and/or morphological distortion. However, both methods are much more rapid than histological preparations with the attendant fixation, embedding in a support medium (e.g., paraffin), sectioning, and staining steps that require considerable time and effort. The brush smear and cytospin techniques both result in a dilute distribution of whole cells, rather than sectioned material, making differential distinctions much easier. Methods to access the marrow cavity of the femur of the mouse are described in Protocol 9. These methods permit ready access to the cells for either direct brush smear or suspension for cytocentrifuge preparation.

Protocol 9

Bone Marrow Biopsy of the Mouse

The recovery of bone marrow for cell analysis by microscopic examination or for any of a number of other purposes such as adoptive cell transfer is best described as a messy task, in part due to the small size of an adult mouse. The bones of the hind legs offer reasonable sites from which marrow can be readily recovered. This protocol describes how to recover the most accessible marrow population in the mouse—from within the shaft of a femur.

MATERIALS

CAUTION: See Appendix for proper handling of materials marked with <!>.

Reagents/Animals

Isopropyl <!> or ethyl <!> alcohol (70%) in a squirt bottle
Mouse from which marrow is to be harvested

Equipment

Dissecting scissors (sharp)
Forceps
Gauze pads
Scalpels (sterile; no. 4 handles with no. 10 blades are recommended)

METHOD

Two procedures are presented here. Steps 1–3 plus Steps 4–7 outline a quicker and less labor-intensive alternative for marrow preparation that limits the dissection of the femur to a point at which the marrow cavity is simply accessible. Steps 1–3 plus Steps 8–13 outline a detail-oriented protocol to recover individual leg bones for counting/cell analysis as well as tissue culture protocols. Regardless of the method of marrow recovery selected, if a cell suspension is required from one of the femora for other experimental purposes, be sure to set it aside first. That is, because a smear may be made from any number of marrow sites, the priority should be the preparation of cell suspensions for these other experimental goals.

Dissection of Femur

1. Sacrifice a mouse by cervical dislocation (or as specified in the approved IACUC protocol) and place the euthanized mouse in dorsal recumbancy.
2. Wet the fur with 70% isopropyl or ethyl alcohol and make a mid-abdominal transverse incision in the skin (see Video 7, Surgical Exposure of Femur).
3. Grasp the skin on each side of the incision using the thumb and forefinger of each gloved hand and simultaneously remove the skin both cephalad and caudally, making sure to expose the upper and lower hind leg musculature.

   Proceed to Step 4 for the quicker procedure or to Step 8 for completion of the excision of the intact femur.

Isolation of Marrow Cavity of Femur

4. Grasp the exposed muscle mass with the forceps. Insert the points of the scissors on one side of the shaft and separate the points, moving the muscle mass away from the bone (see Video 7, Surgical Exposure of Femur). Repeat this procedure on the other side of the shaft.
5. After the femur is relatively free of muscle, use scissors to cut the bone transversely at the distal joint (i.e., the knee). Grasp the femur at the severed tip and cut at the head of the femur as close to the pelvic articulation as possible.
6. Place the femur on gauze and scrape it clean of muscle using a scalpel.
7. Use a scalpel with a clean blade and apply pressure longitudinally to crack the cylinder open lengthwise to expose the contained marrow.

   ▸ *See Troubleshooting. If preparing a brush smear of femoral marrow for cell analysis by microscopic examination, proceed immediately to Protocol 10.*

Complete Isolation of the Femur

8. Use dissection scissors to sever the pelvis on either side of the backbone.
9. Cut or twist as close to the pelvic articulation as possible to free the femur. Be careful not to damage the epiphysis.
10. Remove each leg (including the attached muscle), pulling it free of the skin. Use scissors to sever the feet from the legs at the ankle, again being careful not to damage the epiphyses of the tibias, because these bones may be needed for marrow recovery.
11. Trim away the bulk of the leg muscles with dissection scissors. Separate the muscle-trimmed legs at the knee joint by hyperextension and strip away the remaining muscle of each femur with tissue paper.

12. Once cleaned, remove both epiphyses of the femur by cutting with a scalpel, leaving a femoral "cylinder" with as yet undisturbed marrow in its cavity.
13. Use a scalpel with a clean blade to split the cylinder open lengthwise to expose the contained marrow.

    ▸ *See Troubleshooting. If preparing a brush smear of femoral marrow for cell analysis by microscopic examination, proceed immediately to Protocol 10.*

TROUBLESHOOTING

Problem (Steps 7 or 13): Long bone is split in such a way as to make the marrow cavity inaccessible.

Solution: It may be possible to attempt to split the femur again because enough marrow may be accessible to obtain an adequate sample using the brush smear procedure. Otherwise, access other bones.

DISCUSSION

The differential cellularity of all marrow is equivalent. One marrow shaft is more than adequate to yield enough cells for a brush smear if the contralateral femoral marrow is needed for other purposes. However, if collection of marrow from both femora is necessary for other purposes, one of the tibiae provides ample cells for the preparation of slides. In some instances, however, both femora and tibiae are required for cell suspensions needed for adoptive cell transplants, FACS (fluorescence-activated cell sorting) analysis, or other procedures. In this event, either humerus is sufficient. It is also possible to obtain marrow samples from either the sternum or the iliac crest using similar procedures to expose the marrow-containing cavities of these sites. It is important to note, however, that extramedullary sites of hematopoiesis such as the spleen differ considerably from marrow in their differential cellular composition.

A similar stepwise process may be followed for preparing a marrow smear from other bones—the bone is excised and the marrow is accessed. The humerus can present challenges because it is relatively small and is surrounded by a considerable muscle mass. Procedures for accessing sternal or pelvic marrow will also be similar in character. Irrespective of the marrow source, it is essential only that the marrow be exposed sufficiently to permit access with a wetted brush tip (see Protocol 10).

Protocol 10

Preparation of a Bone Marrow Brush Smear

This protocol describes the preparation of a slide by brush smear of femoral marrow for cell differential analysis. The technique involves the use of a small brush—often used as a preferred instrument by watercolor painters—to create a smear of marrow cells on a glass slide. The marrow cells are mixed with slightly hypotonic serum, so that the cells retain their normal dimensions as the smears dry.

MATERIALS

Reagents/Animals

Hypotonic fetal calf serum (two parts serum to one part distilled H_2O)

It is easiest to prepare 3–10 mL of hypotonic fetal calf serum and distribute it into 1.5-mL microcentrifuge tubes (~0.25 mL/tube) before use and freeze (–20°C). The necessary number of tubes may then be retrieved and thawed just before use. Normally, only one tube is needed for a group of animals, but it is best to have at least two available in the event that one gets contaminated by the reinsertion of a used brush into the tube.

Shaft of mouse femur (isolated as described in Protocol 9)

Equipment

Camel hair brush (5/O)

One brush is required for every animal from which marrow smears will be taken. The best results are obtained with a 5/O camel hair brush. Other brush styles and sizes may be used as well. We have used 3/O and 4/O brushes, and some have been made of materials other than camel hair. However, these brushes have been found to produce less favorable results. Although expensive, 5/O camel hair brushes provide the best results and may be used repeatedly if cleaned with a nonionic detergent (see Steps 6–8). Although camel hair brushes may have originally been made with the fine hairs of a camel, these brushes are now more commonly made from the tail hairs of squirrels or even the ear hairs of bovine calves.

Containers with dilute nonionic (i.e., mild) soap, tap H_2O, and distilled water (for washing brushes; see Steps 6–8)

Coverslips

The coverslip must be large enough to contain the entire length of the smear; we usually use 24 x 50-mm 1.5-oz. slips.

Gauze (2 x 2 or 4 x 4 inches) or other absorbent towels
Glass slides (washed and prepared as described in Protocol 5)
Use two slides for each animal.

METHOD

Preparation of a Brush Smear

1. Wet a brush in slightly hypotonic calf serum. Remove excess serum by quickly drawing the brush across an absorbent surface (gauze or towel) to form a point with the bristles. Be careful not to remove all the fluid.
2. Place the pointed tip of the wetted brush into the marrow cavity and mix the hypotonic serum with the marrow to create a "slurry." Do not include the entire contents of the selected long bone, only enough to make the slurry sufficient to wet the bristles of the brush.

   This step requires some practice, and the appropriate consistency will become apparent through a process of trial and error.
3. Place a droplet of the mixture from the brush onto a glass slide near the frosted end and "paint" the suspended marrow cells from that source droplet using a steady motion (see Video 8, Brush Smear of Marrow).

   The cells are not being physically "brushed" and/or "painted" onto the slide and thus very little pressure is needed. Rather, the brush is a vehicle by which the marrow cell suspension is spread onto the glass to generate numerous areas from which single cells may be identified. When performing this procedure, the bristles of the brush should be only slightly bent to avoid creating a smear that is too thick.
4. Place two to four parallel rows on the slide by repeating the smear with material already on the brush or from the droplet on the slide or by retrieving more from the marrow slurry.
5. Let the smears air-dry completely before proceeding to fix, stain, mount, and visualize them as described for blood films in Protocol 7. When mounting, use coverslips that are large enough to contain the entire length of the smear.

   ▸ *See Troubleshooting.*

Washing Brushes

6. Soak the brushes for ~1 min in a container of a dilute nonionic (i.e., mild) soap.
7. Rinse the washed brushes with copious amounts of tap water and then with distilled H_2O.

   It is critical to rinse the brushes free of soap before reusing the brushes, because the residual soap will lyse cells.
8. Air-dry the brushes.

TROUBLESHOOTING

Problem (Step 5): Coverslip does not rest evenly on the slide.

Solution: Examine the preparation for boney spicules or other contaminating debris; remove some of this material with forceps and recoverslip the slide. In addition, if the smear is too thick in spots, the coverslip will not lie flat. This may be avoided at the time of preparation if care is taken to make suitably thin streaks with a sufficiently diluted marrow slurry.

Problem (Step 5): Bone marrow brush smears are too thick.

Solution: The smear brush may be too dry when initiating the creation of the marrow slurry. It may be necessary to work more rapidly or make the slurry more dilute with hypotonic serum. If this is necessary, make sure not to return a "marrow-laden" brush to the hypotonic serum reservoir. Use a fresh brush wetted with the serum to dilute the marrow slurry in situ. Alternatively, the smear brush may be loaded with too much marrow slurry, in which case, you can reduce the bulk of material on the brush. Swipe the bristles with a KimWipe and return the brush to the marrow to get a smaller amount on the brush. Otherwise, use a new brush to access the marrow, loading it with fresh slurry. Practice will lead to the use of better volumes on the initial effort.

Problem (Step 5): Too many cells are broken and nuclear material is streaked in the stained preparation.

Solution: Perhaps too much pressure was used on the brush stroke during application of the marrow slurry to the slide. This may be remedied by ensuring that the bristles of the brush are only slightly bent, drawing out the slurry droplet.

Problem (Step 5): Bone marrow brush smears are too thin.

Solution: The smear brush may be too wet when initiating the creation of the marrow slurry. If this is the case, reduce the bulk on the brush by swiping it with a KimWipe and return the brush to the marrow and to obtain a lesser amount of the slurry on the brush. Practice will lead to the use of better volumes on the initial effort. The smear brush may be loaded with insufficient marrow slurry at the start of the slide preparation. It is possible to return to the marrow with the brush and "reload" it with more marrow slurry. A brush stroke too light when applying the marrow slurry to the slide may be remedied by paying attention to the bristles of the brush during application of the smear to the slide, ensuring that they are slightly bent to draw out the slurry droplet.

Problem (Step 5): Cells are shriveled.

Solution: Ensure that the hypotonic serum was prepared with a 2:1 ratio of fetal calf serum:H_2O. Because the serum is slightly hypotonic, the cells retain their normal dimensions as the smears dry. However, if the smear is too thick or dries too slowly, the serum will become hypertonic with the evaporation of water, and the cells will shrivel and be impossible to type. If this occurs, use a clean brush to obtain additional material. Returning a used brush tip to the hypotonic serum reservoir contaminates it, making it useless for subsequent marrow collections.

DISCUSSION

A similar stepwise process may be followed for preparing marrow brush smears from bones other than the femur. Irrespective of the marrow source, it is essential that only the marrow be exposed sufficiently to permit access with a wetted brush tip. Once this is accomplished, it is a simple matter to prepare the slurry in situ and proceed with applying the cells to the slide.

Similar to the preparation of a smear from the cavity of one of the long bones, a smear preparation from other lymphohematopoietic tissues (e.g., spleen, liver, lymph node, thymus, or fetal tissues) is accomplished by first making a clean cut on the surface of the target tissue. A cell suspension is subsequently prepared by using a hypotonic serum-wetted 5/O brush to make a slight slurry at that cut surface. Using the brush, the cells of the serum suspension are transferred onto the slides in the same manner as described for bone marrow. However, with this technique, care must be taken to avoid preparing a smear with a high density of cells. Practice will allow refinement of the procedure and ultimate success. Despite the additional effort associated with this technique, the smear strategy is preferable to making simple “touch imprints” on slides using the freshly cut surface of a tissue, partly because of the quality and representative character of the leukocytes obtained.

Protocol 11

Preparation of a Bone Marrow Suspension for Cytospin Analysis

Certainly, the most common (although not necessarily the most preferable) method of cell preparation for microscopy is by cytocentrifugation (i.e., a cytospin). The cytospin preparation of femoral marrow cells is a relatively straightforward exercise and may be particularly useful if marrow cells are being prepared for another procedure, such as with in vitro tests, FACS analysis, or adoptive cell transfer. Then, instead of preparing smears as described in Protocol 10, it is simpler to access an aliquot of the suspension and prepare slides for microscopic examination by cytocentrifugation. However, although convenient, the quality of slide preparation for microscopy by this method is significantly poorer relative to preparation by the brush smear technique.

MATERIALS

Reagents/Animals

Leukocyte-suspending solution

Use a balanced salt solution such as Dulbecco's PBS (phosphate-buffered saline) or a minimal essential medium such as Hank's or Eagle's, warmed to 37°C.

Saline solution containing 5%–10% fetal calf serum for wetting slides before cytocentrifugation (see Video 6, Cytocentrifuge Procedures)

Shaft of mouse femur (isolated as described in Protocol 9)

Equipment

Bucket with ice

Conical centrifuge tube (15 mL, sterile)

Cytocentrifuge and cytospin slide apparatus (see Video 6, Cytocentrifuge Procedures)

Disposable transfer pipette (sterile)

Syringe (3 mL) with either a 25- or 26-gauge needle

METHOD

If sterility is of concern, precautions must be taken to ensure that all instruments, tubes, and solutions are sterile. All procedures are performed in a sterile field in a laminar flow biosafety cabinet.

1. Fit a 3-mL syringe with either a 25- or 26-gauge needle, depending on the size of the diameter of the shaft of the bone to be flushed. Fill the syringe with leukocyte-suspending solution warmed to 37°C.
2. Insert the needle into one end of the bone to be flushed and express the suspending solution through the marrow cavity to collect it, along with the marrow "plug," in a sterile 15-mL conical centrifuge tube.
3. Disassociate the marrow plug into single cells by repeated aspiration (20–30 times) with a sterile disposal transfer pipette. Keep the suspension on ice during this procedure.

   In general, a femur from an otherwise healthy mouse of either gender whose weight is ~20–25 g will contain 18×10^6 to 22×10^6 nucleated cells. The suspension, if prepared in ~3 mL, will have ~7×10^6 cells/mL.
4. Dilute the suspension in leukocyte-suspending solution so that a volume of 0.10–0.12 mL (100–120 µL) contains ~30,000 cells for cytospin preparation.

   The estimated number of cells per femur may be used to guide the dilution steps, or a hemacytometer count may be performed as described in Protocol 4.
5. Perform cytocentrifugation (i.e., cytospin) as shown in Video 6, Cytocentrifuge Procedures, with two slides per sample. Adhere to the manufacturer's operation manual.

   Put some effort into defining the technical parameters of the cytospin device. Variables such as the speed and duration of the spin, as well as the initial volume of cell suspension, should be maximized for the different cell populations being studied. Only in this way can one confidently report reproducible results from this technique.
6. Fix, stain, mount, and visualize the cells as described for a peripheral blood film (see Protocol 7).

   ▸ *See Troubleshooting.*

TROUBLESHOOTING

Problem (Step 6): Too few or too many leukocytes are present on the cytospin preparation. If there are too few cells, it may be difficult to find the cells and the number of lysed or fragmented cells will increase significantly. If there are too many cells, the overlapping and crowding may change the morphology of the cells as well as their staining properties.

Solution: Review the calculations and/or cell counts used to prepare the cell suspension for cytocentrifugation. The number of cells suggested for a cytospin varies greatly with "reliable" sources, which suggest as little as a few thousand to as many as 100,000 cells. In our experience, the window for the appropriate number of cells is much smaller, with a focus on 30,000 cells.

6 Cell Differential Assessments of Bone Marrow

THE EXAMINATION AND CLASSIFICATION of hematopoietically active bone marrow are at best a daunting task. Not only are the lineages intermixed, but in most cases, the early progenitors of one lineage are nearly indistinguishable from those of the other lineages. Moreover, the cells are not stratified as in many other proliferative tissues (e.g., skin or intestine). In addition, because many of the cells are mitotically active at various stages of the cell cycle, their DNA content and overall sizes are variable; thus, the degree of chromosomal organization for mitotic events adds a measure of complexity to the task of classification based on morphology. This chapter presents a stepwise strategy to differentiate the hematopoietic cellular elements of the marrow. In this way, it should be possible to acquire sufficient skill to accomplish a differential assessment of the cellular composition of hematopoietic tissue with confidence.

GENERAL STRATEGY FOR ASSESSING SMEARS

As with the assessment of a peripheral blood film (Chapter 4), the marrow smear should first be screened using low-power magnification (e.g., 10x objective with 10x oculars). The smear may resemble a "ribbon" of cells because of the use of a brush during preparation (see Protocol 10). Normally, cell distributions range from tightly clumped (i.e., with virtually impossible to identify individual cells) to nicely separated fields (Fig. 12). Only areas of well-separated cells that are clearly stained should be selected for examination at a higher magnification. Streaks of purple material (DNA from disrupted cells) and denuded cells (nuclei without cytoplasm) will more than likely be seen (Fig. 13). If these events comprise ≥5% (~1 in 20) of a given microscopic field, another area should be selected. However, if these events are characteristic of the entire preparation, there may be something wrong with the procedure or it may be pathognomic of a condition in the animal.

A

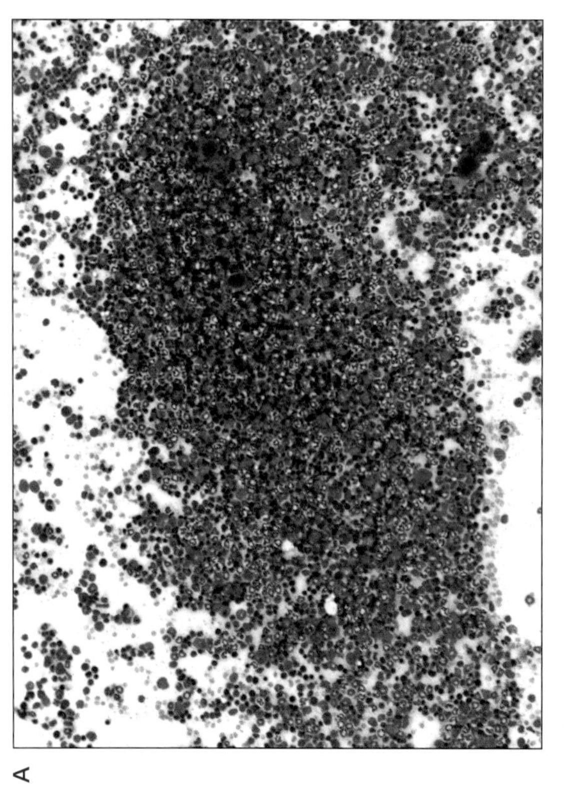

B

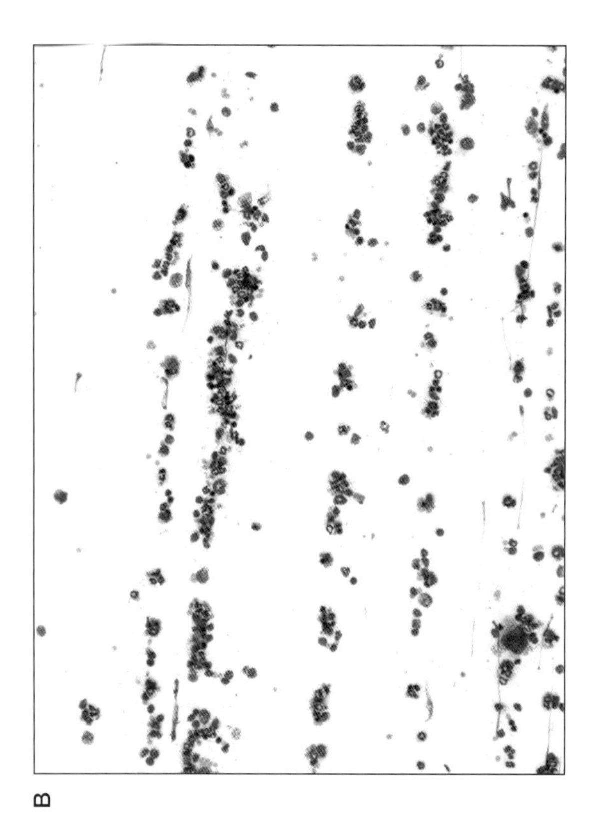

C

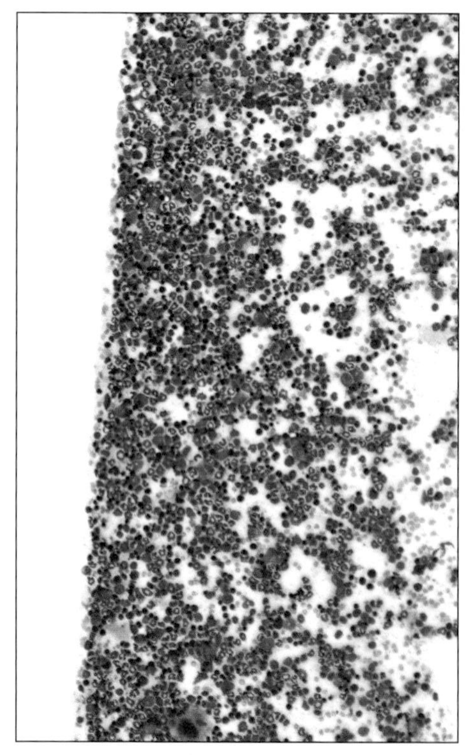

FIGURE 12. Microscopic views of features characteristic of typical brush smears of femoral marrow. (*A*) Regions of dense and/or "clumped" areas of marrow in a smear. These areas are not appropriate for microscopic examination and cell differential assessment due to crowding and accompanying changes in cellular morphology as well as staining due to "neighborhood" effects. (*B,C*) Two areas of a typical bone marrow smear with varied cell densities that are ideal for cell differential assessment. At still higher magnifications, contiguous fields will have cells sufficiently distinct from one another so as to allow accurate lineage determination.

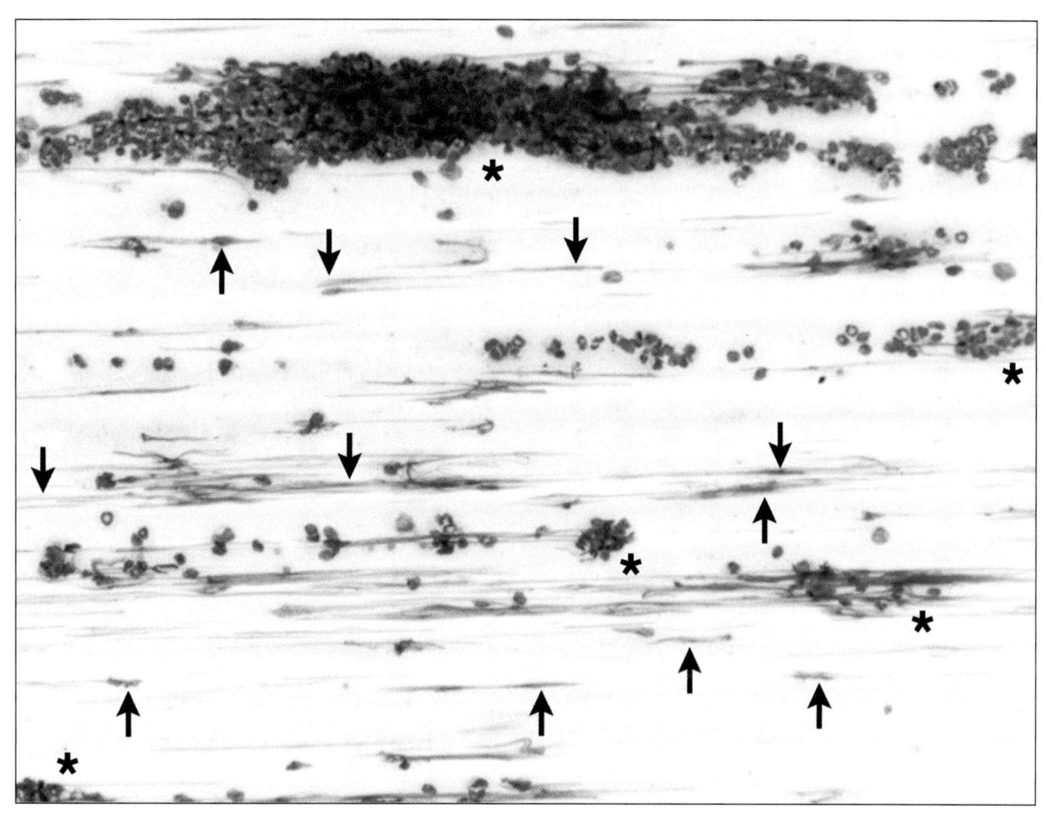

FIGURE 13. Denuded cells and dense clumps of cells in a bone marrow smear are areas of concern and should be avoided during cell differential assessments. A coarse motion with the brush (e.g., excessive brush pressure against the slide) may result in clusters and/or nests of cells (*) that are unidentifiable and an inordinate number of broken cells of which the only remaining substance is residual nuclear material (arrows). Occasionally, residual granules or other cellular structures may also be evident adjacent to "naked" nuclei. The telltale presence of such cellular remnants should not be used to extrapolate the identity of the "lost" cell because such guess work is just that and will invariably lead to a skewed count. Bottom line: ≥95% of the cells in a field for which a count is being taken must be identifiable for that area to be considered.

Before any lineage assignments are made, it is necessary to first establish that the smear is representative and that cells are evenly distributed. A sufficient number of cells within several regions should be identifiable to eliminate the potential prejudice that can occur as one scans a slide to select a single "favorable" area. To avoid this, an area of the slide with several adjacent fields of suitably spread cells should be identified using low-power magnification. These areas will provide the necessary numbers of cells to allow accrued counts within the selected and adjacent fields. The accumulation of a differential count of ≥200 cells is a minimal standard and contiguous fields should be used, if possible. The number of nucleated cells within a marrow smear is considerably greater than that of a peripheral blood film, making it much easier to accumulate a ≥200 cell count without extensively searching the preparation and thus facilitating the use of higher magnifications. The cell count being accumulated will depend on the nature of the study and the frequency of the cell(s) of interest. Generally, a 200-cell count is sufficient as an initial assessment.

It may be possible to accomplish differential classification using a total magnification of 400X (a 40X objective with 10X oculars). However, the task is somewhat easier with the added resolution available with objectives of yet greater magnification. A 63X objective with 10X oculars combines for a useful magnification with added resolving capability. On occasion, it may be necessary to resort to a 100X objective to appreciate the subtle features and accurately define the lineage or stage of some cells.

Technical Tip: Although a 100x objective with 10x oculars (1000x total) provides a significant increase in magnification relative to 63x objectives with 10x oculars (630x total), there is not a similar increase in resolving power (0.24 µm vs. 0.26 µm, respectively). The subtleties of using intracellular morphologies as a tool for cell differential analysis are highly dependent on an objective's resolving capability and not simply its total magnification. Bottom line: The highest available power is not necessarily better because of diminished light reaching the oculars and only a marginal increase in resolution, i.e., in most cases, a 63x objective with 10x oculars will provide the best overall image for cell differential analysis.

The photomicrographs herein that illustrate the various cell types are useful, although it is important to remember that they are nonetheless images and are necessarily two dimensional. In other words, it is whole cells that are being placed on the glass surface and, although they appear to be flat, they actually have the full complement of their composition available for microscopic examination. Smears and films should always be examined by constantly moving the objective up and down through the plane of focus. Only in this way is it possible to appreciate all of the structural subtleties of the cells.

AN OVERVIEW OF THE CELLULAR COMPOSITION OF MARROW

The marrow of an unmanipulated and otherwise healthy mouse, regardless of strain, gender, postweaning age, housing standard, and bone site, will typically be composed of mature and developing cells (Table 3). Although attention must be paid to the entire cellular composition of the marrow smear, practically speaking, marrow cell differentials usually focus

TABLE 3. Nomenclature for morphologically identifiable hematopoietic lineages in the mouse

Erythroid cells
*Erythroblast–proerythroblast–basophilic erythroblast–polychromatic erythroblast–orthochromatic erythroblast–reticulocyte–**erythrocyte (RBC)***

Polymorphonuclear granulocytes: neutrophils (i.e., PMNs) and eosinophils
*Myeloblast–promyelocyte–myelocyte–metamyelocyte–**neutrophil (Neu)/eosinophil (Eos)***

Lymphocytes
*Lymphoblast–prolymphocyte–**lymphocyte (Lym)***

Plasma cells (B cells)
*Plasmablast–proplasma cell (pro-B cell)–**plasma cell (PC)***

Monocytes
*Monoblast–promonocyte–**monocyte (Mo)***

Megakaryocytes
*Megakaryoblast–**megakaryocyte (Meg)***

on two granulocytic lineages (neutrophils and eosinophils) and cells of the erythroid, lymphoid, and monocytic series. Nucleated cells of these lineages will represent >99% of the marrow cellularity. Megakaryocytes may also be noted, although normally they do not comprise >0.1% of marrow. If their numbers are increased, this should be noted and assessed accordingly. Similarly, basophil granulocytes are rare cells and generally not considered in differential assessments.

1. **Erythroid cells:** These are the hemoglobinizing mononuclear cells that terminate, as a series, in cells that extrude their nuclei resulting in anucleate erythrocytes (i.e., RBCs). Although not "leukocytes," the erythroid progenitors are nucleated and constitute a significant component of hematopoietic tissues.
2. **Polymorphonuclear granulocytes (PMNs):** The nuclei of this subtype of leukocytes are convoluted and polylobular and may present with a "ring" shape. Hence, there are multiple (*poly*) shapes (*morpho*) and levels of chromatin condensation for the nuclei (*nuclear*) of these cells (PMNs).

   Eosinophils and *neutrophils* are PMNs with prominent cytoplasmic granulation (thus the name *granulocytes*). These granules are either basic (i.e., acid-loving or acidophilic) and stain magenta/orange/pink with the acid dye eosin (eosinophils) or pH neutral and are unaffected by the stains methylene blue or eosin (neutrophils). Occasionally, one may encounter a granulocyte with cytoplasmic granules exhibiting extreme basophilia. The incidence of these cells (basophils) is sufficiently infrequent so as to not affect consideration in a typical differential. An incidence greater than 1/100 may be reflective of pathology and is certainly to be considered abnormal.

Did you know? The term polymorphonuclear granulocyte (i.e., PMN) is now currently used in the literature as a description of, and an alternative name for, neutrophils. This is particularly true in studies describing human cells, but it has since carried over to most reports assessing mouse models of human disease. Bottom line: Although eosinophils certainly can be described as PMNs, by existing convention, they are generally no longer described as such, leaving PMN synonymous for neutrophils.

3. **Lymphomononuclear cells**
   - *Lymphoid cells:* These are B and T lymphocytes of varied maturations, including a unique antibody-secreting population of B lymphocytes also known as *plasma cells.*
   - *Monocytes:* Cells of the monocyte series lead to circulating monocytes and, in turn, tissue-resident macrophages (e.g., Kupffer cells [liver], microglial cells [brain], and alveolar [lung] and splenic macrophages).
   - *Megakaryocytes:* These are extraordinarily large platelet-producing cells limited exclusively to hematopoietically active compartments.

The nomenclature for the various intermediate stages of each hematopoietic lineage has been a matter of debate for decades. If one compares these discussions with phenotypic delineations based on cell-surface markers, it is not too difficult to appreciate the challenge. The binding of monoclonal antibodies to define cluster of diffferentiation (CD) cell-surface expression (CD^+ or CD^-)—as well as the identification of cells based on markers characteristic of mature cell lineages (Lin^+ or Lin^-) or the elevated/depressed expression of a given CD marker (CD^{hi} or CD^{lo})—is not without ambiguity. Because the differentiation process is dynamic and not synchronous, categories of cells are not distinct and cell lineage distinctions are necessarily somewhat arbitrary. However, these definitions can be very useful to understand the level of maturity within a lineage (McGarry and Stewart 1991). For the following discussion on the identity and characteristics of each hematopoietic lineage, see the classification categories listed in Table 3.

Finally, it is noteworthy that in addition to mature and developing cells classically defined as RBCs and WBCs, other cells will be encountered that may be confusing or at the least elicit a pause in one's assessments. These miscellaneous marrow cells may include some related to bone metabolism (e.g., osteoclasts and osteoblasts), cells associated with the structural integrity of the marrow (e.g., fibroblasts and blood vessel endothelial cells), and very primitive hematopoietic stem cells and/or undefined progenitor cells. However, these normally comprise a very small percentage of the marrow cellular complement (<1% of total cells examined) and are relevant only as a particular experimental objective may dictate. In addition, some marrow cells may exhibit phagocytosis. These are functional cells of the marrow and are not immediately components of hematopoiesis, i.e., they have no immediate significance related to the cell differential assessment of the hematopoietic marrow compartment.

> ***Technical Tip:*** Because a marrow differential is composed of only identifiable cells, caution should be exercised if the proportion of "unidentifiable" cells of a given slide preparation exceeds 10%. This is especially important because ignoring these "lost cells" may mask marrow lineage or cell maturity shifts, significant pathologies in the experimental animals under study, or technical errors associated with the bone marrow preparation.

Even if the smear is well prepared, with cells sufficiently separated, the identification of each of the cells in a field is challenging to the experienced examiner as well as the novice. Start marrow examination by finding representative examples of select "hallmark" lineages (Fig. 14). Find *megakaryocytes*, the series of nucleated immature cells from which platelets are derived (Fig. 14A); *plasma cells*, characterized as end-stage antibody-producing cells (Fig. 14B); and *eosinophils*, granulocytes characterized by both polymorphonuclear changes and eosin-staining cytoplasmic granules (Fig. 14C).

> ***Technical Tip:*** Mature cells of each hematopoietic cell lineage have distinct features that are excellent for referencing the expected nuclear and cytoplasmic morphologies of developing hematopoietic populations as well as the quality of the marrow smear and its staining.

Mature cells of these three lineages are not easily confused with other cells of hematopoietic marrow. The time spent identifying these cells will help to familiarize the investigator with the variability of cellular morphologies within somewhat easily defined lineages. In addition, it is necessary to seek out these cells to help to define the parameters of stain and fixation, which change with time and use. These skills, once acquired, will contribute to building experience for the more challenging definitions of the other hematopoietic cells of the marrow.

The following discussion is intended to serve as a "primer" for the microscopic examination of Romanowsky-dye-stained marrow and, in turn, peripheral blood cell preparations (see Chapter 4). The initial strategy to become acquainted with the various cellular elements of hematopoiesis rests on the ability to identify the most recognizable cell (most mature and characteristic) within an individual lineage. That is, seek cells that are obviously related to a mature cell of a given lineage (see the poster, Mouse Peripheral Blood Cells) but that also exhibit less terminally differentiated qualities, ignoring for the moment cells of all other cell lineages. The objective is to recognize common morphological features of a single lineage, identifying cells of *decreasing* maturity until a "next-least-mature cell" is no longer identifiable as lineage specific. Maturing cells have vari-

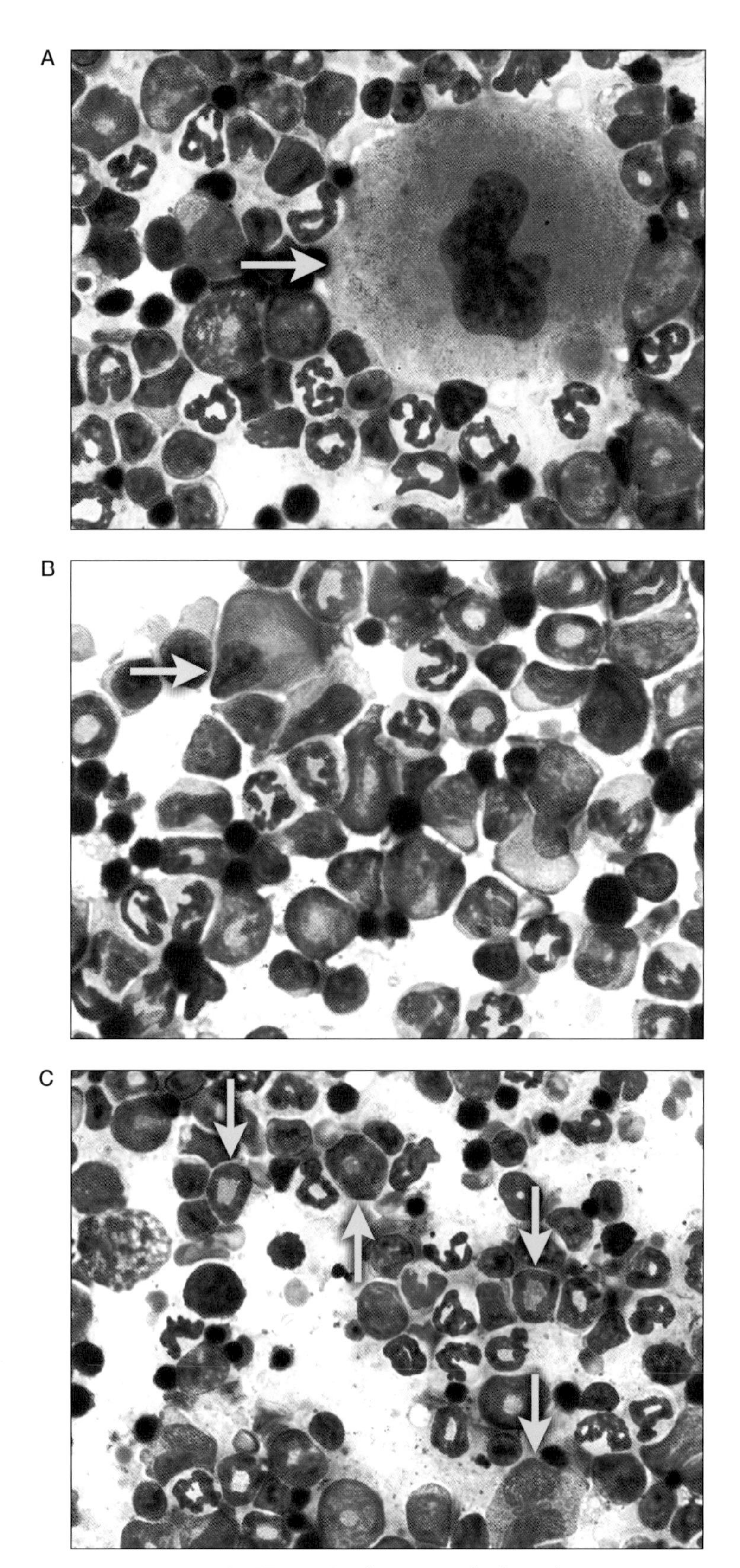

FIGURE 14. *(See facing page for legend.)*

ably blue cytoplasm—this is a characteristic of cells actively engaged in protein synthesis. These cells are basophilic, owing to elevated cytoplasmic RNA levels. In addition, immature, developing cells (with the exception of megakaryocytes) are still replicating (undergoing karyokinesis *and* cytokinesis) and therefore are duplicating DNA from 2N to 4N (i.e., exhibiting chromosomal structural changes and cell morphologies related to the mitotic process itself). Moreover, the nuclear-to-cytoplasmic ratio decreases with maturity in all cell lineages; the earliest of cells are almost all nucleus, bearing only the scantest rim of cytoplasm.

Technical Tip: The least-mature cells of the neutrophil/eosinophil, monocytic, erythroid, and lymphoid series are at best difficult to distinguish from one another and at worst simply indistinguishable.

Finally, when assessing smears from multiple organs, it is necessary to keep in mind that the differential cellularity of marrow from all bones is equivalent. However, this is not true of extramedullary sites of hematopoiesis such as the spleen, which differ considerably from marrow in their differential cellular composition.

ERYTHROPOIETIC SERIES

Mature cells of the erythroid series (i.e., *erythrocytes* or RBCs) are easy to distinguish: They are anucleate biconcave disks containing hemoglobin and thus have the familiar red color. Their nuclei are lost (extruded) in the terminal stages of differentiation. The progressive stages of morphological maturation are best identified in reverse (i.e., going from most to least mature) (Fig. 15). Residual cytoplasmic RNA required for the synthesis of the hemoglobin will cast gray-blue shading on the most recently formed erythrocytes, termed *reticulocytes.* Reticulocytes are anucleate cells that will normally are found both within erythropoietic spaces and in the circulation (see Chapter 4). Cells that give rise to anucleate reticulocytes have condensed, heterochromatic nuclei and bluish-gray, homogeneous-staining cytoplasm. These cells are known as *orthochromatic erythroblasts.* On occasion, one may even find a reticulocyte with a partially extruded nucleus.

A distinctly less-mature cell of this series will have a similar nucleus, but the cytoplasm will be more slate blue and "patchy." The nucleus will also be less heterochromatic and may not be postmitotic. This cell is iden-

FIGURE 14. Photomicrographs of hallmark cells found within mouse hematopoietic tissues. Before beginning a differential assessment of hematopoietic tissues such as femoral bone marrow or spleen, it is beneficial to locate and consider the staining features of three hallmark leukocytic cell types: (*A*) megakaryocytes, (*B*) plasma cells, and (*C*) eosinophils. The identified cells of each panel are denoted with arrows.

A

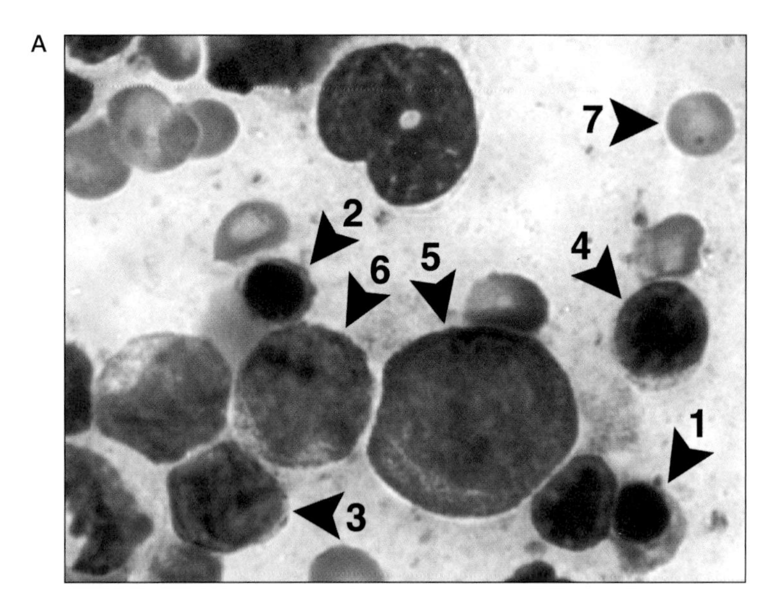

B

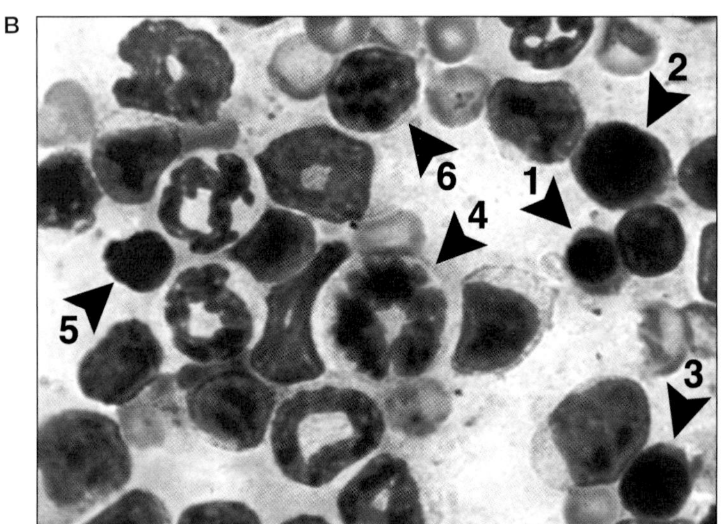

FIGURE 15. Differential assessment of the mouse erythroid lineage. Erythrocytes, or mature anucleate RBCs, are present in marrow (panels *A*[7], *D*[6], and *F*[2]). Reticulocytes are those cells that are anucleate, fully hemoglobinized, yet due to their young age, still retain some trace globin RNA, giving them a blue-gray hue (panels *D*[5] and *E*[3]). The orthochromatic erythroblast has a nucleus, albeit highly condensed before extrusion. The cytoplasm is strongly similar to that of the reticulocyte, at least in patches (panels *D*[4], *E*[2], *E*[5], and *F*[1]). The polychromatic erythroblast has a less-condensed nucleus and more basophilic cytoplasm with patchy pale areas of hemoglobinization (panels *A*[2] and *B*[1]). Basophilic erythroblasts have nuclei that are active mitotically and cytoplasm that is intensely blue (panels *D*[2] and *D*[3]). Distinguishing features of the proerythroblast are best defined within the context of neighbors, i.e., their spherical nuclei are more intensely stained and the scant cytoplasm is more intensely blue than similar cells in the vicinity (panels *A*[3], *A*[4], *B*[2], and C[6]). Note that in brush smears of marrow from normal mice, proerythroblasts will most frequently be found in groups. Examples of other cells of interest:

- Panel *A*: (*1*) orthochromatic erythroblast; (*5*) large erythroblast; (*6*) myeloblast
- Panel *B*: (*3*) polychromatic erythroblast; (*4*) neutrophilic myelocyte in metaphase; (*5*) erythroblast; (*6*) erythroblast in early prophase

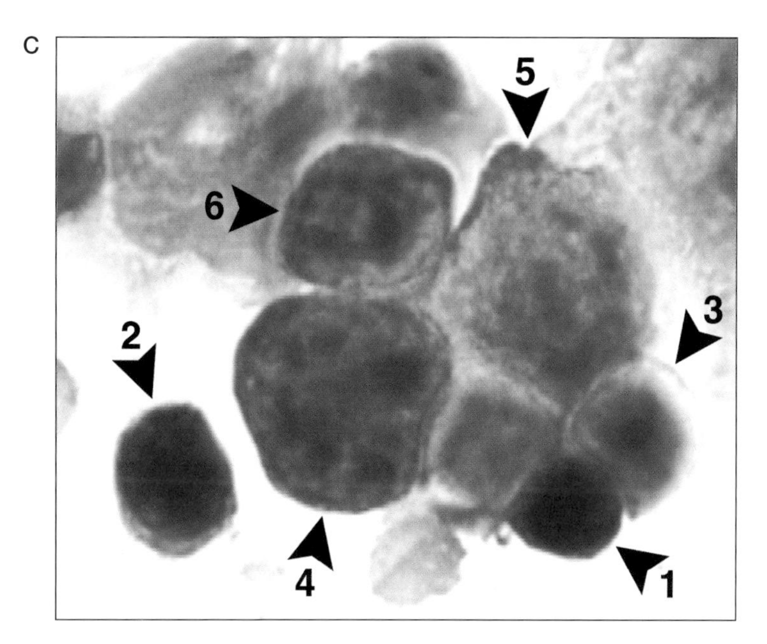

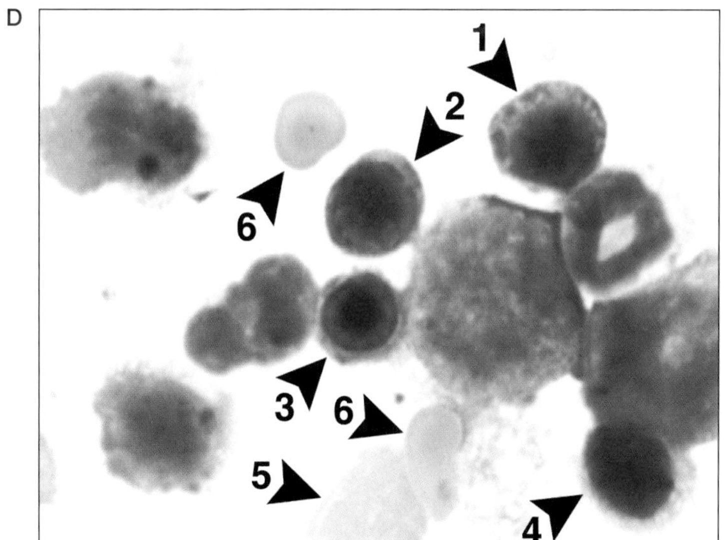

FIGURE 15. *(Continued)*

- Panel *C*: (*1*,*2*) erythroblasts; (*3*) lymphocyte; (*4*) undifferentiated blast cell of an indeterminate lineage with several nucleoli; (*5*) neutrophilic promyelocyte
- Panel *D*: (*1*) polychromatic-to-basophilic erythroblast (intermediate features)

(Continued on following page)

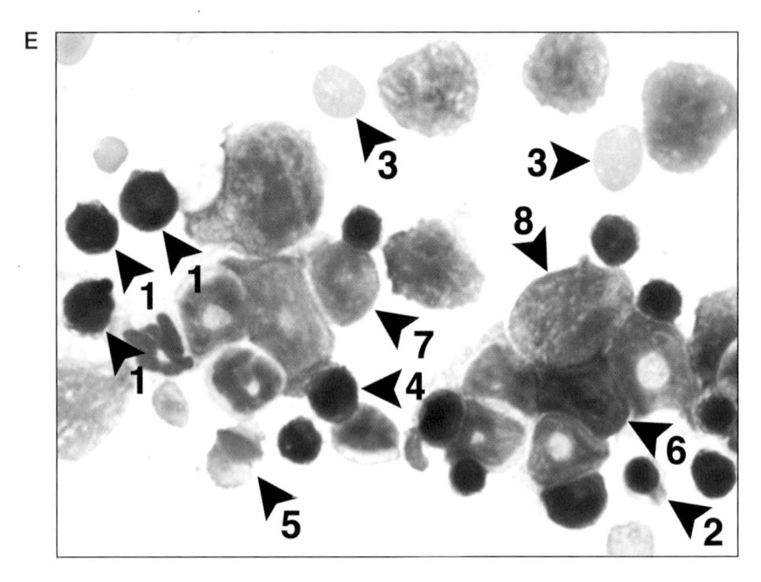

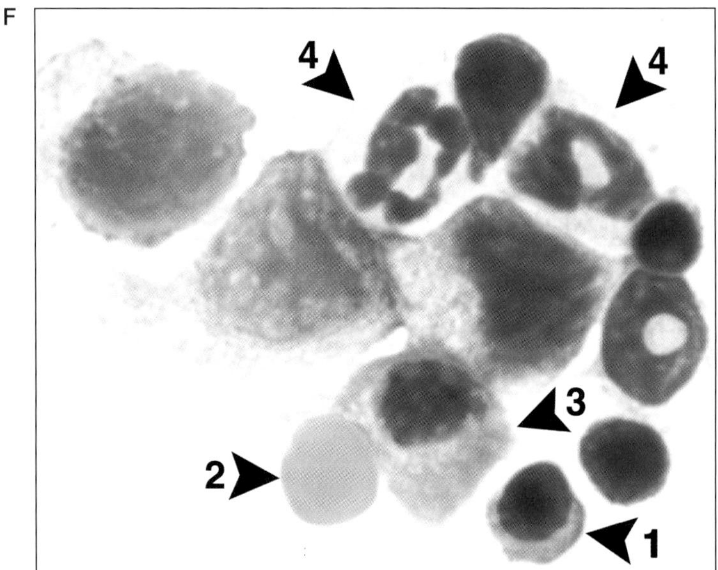

FIGURE 15. *(Continued)*

- Panel *E*: (*1,4*) erythroblasts; (*7*) lymphocyte; (*6*) plasma cell; (*8*) late eosinophil promyelocyte
- Panel *F*: (*3*) plasma cell; (*4*) neutrophil metamyelocyte

tified as a *polychromatic* erythroblast, named as such because of cytoplasmic staining variations from the hemoglobin and RNA. The *basophilic* erythroblast is a precursor to the polychromatic erythroblast and has little to no hemoglobin but an abundance of RNA, conferring an intense cytoplasmic basophilia (i.e., blue hue). The cytoplasm is also "grainy," limited, and may have clear patches. The nucleus of this cell is spherical, textured, and decidedly in various stages of proliferation. The two classifications of erythro-precursor cells that are less mature than the basophilic erythroblast are the *proerythroblast* and the least-mature erythroblast.

Distinguishing between the proerythroblasts and the erythroblasts, and for that matter between the "transition stages" of any of the sequential maturing pairs, is likely to be based on more of a hunch than on an objective fact. Definitions should be associated with "neighbor references" and expected frequencies of prevalence; i.e., one should expect that in the hematopoietic tissue of an otherwise healthy mouse, there will be a progressive increase in the numbers of more mature cells. Thus, given the interplay between proliferation and maturation, there should be more polychromatic erythroblasts than basophilic erythroblasts, and more of those than proerythroblasts, and so forth. Challenges and difficulties occur when one is trying to identify the earliest cells of this lineage (and indeed the earliest developing cells of all the hematopoietic lineages). Fortunately, these indistinguishable progenitors do not comprise a large proportion of the marrow of an otherwise healthy mouse. Certainly, stem cells and the earliest undefinable progenitors will represent fewer than 5% and more likely <1% of any of the lineages discussed.

Technical Tip: The identification of intralineage distinctions is not as relevant as interlineage distinctions. For example, it is more important to recognize differences between polychromatic erythroblasts and cells of similar maturity in the granulocytic or lymphomononuclear lineages than unambiguously segregating erythroblast subtypes.

POLYMORPHONUCLEAR GRANULOCYTES

Eosinophil Granulocytes

Cells of the *eosinophil* lineage may be the most useful for the process of parameter standardization and for appreciating morphological variation in developing cells. Specifically, mature and lineage-committed precursors in this series are unequivocally "marked" by the presence of distinctive eosinophilic secondary granules and unique nuclear morphologies. Thus, the cells of the eosinophil granulocyteic series are easily arrayed to illustrate the entire range of morphologies of proliferating/differentiating hematopoietic cells. The various cells that comprise the maturing eosinophil series are depicted and described in Figure 16.

The most mature subpopulation, *eosinophilic metamyelocytes,* resides in the marrow until the individual cells are released into circulation. These cells are generally no longer proliferative—they have lost all cytoplasmic basophilia (blue hue) and display significant cytoplasmic granulation dominated by eosin-staining (magenta/orange/pink) secondary granules. Eosinophilic metamyelocytes will also have a distinctively "ringed" nucleus with moderate heterochomatic condensation (Fig. 16, D[2] and E[3]). This

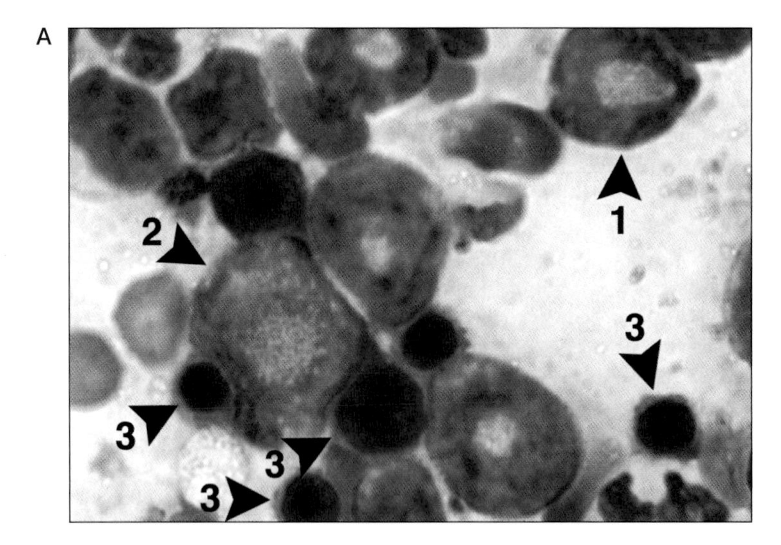

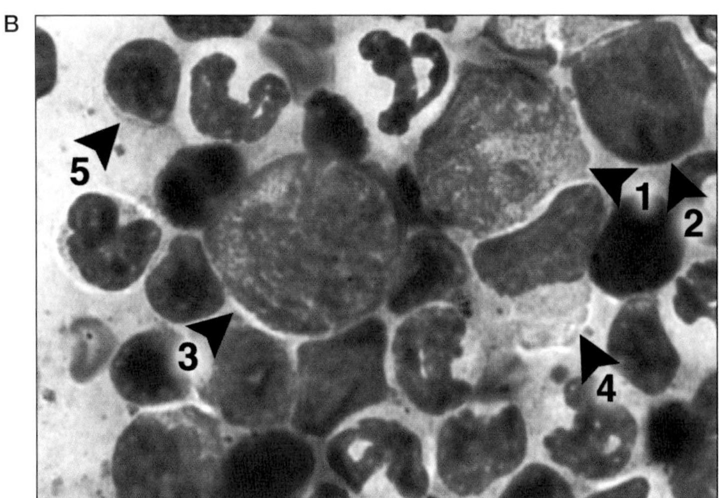

FIGURE 16. Differential assessment of the mouse eosinophil lineage. More mature eosinophils (i.e., late eosinophilic myelocytes and eosinophilic metamyelocytes) have ringed nuclei and cytoplasmic areas with little residual basophilia (i.e., blue hue) as well as acidophilic (magenta/orange/pink) granulation (panels *A*[1], *D*[2], *D*[3], and *E*[3]). The least-mature definitive eosinophils (late eosinophilic promyelocytes/myelocytes) have scant, although definitive, acidophilic (magenta/orange/pink) granulation (panels *E*[1], *E*[2], *B*[1], and *C*[1]). As these cells mature, nuclear indentation and perforation become more pronounced and there is a substantial accumulation of specific cytoplasmic granulation associated with the mature cells of this lineage (see, e.g., panel *A*[2]). Panel *D*(1) is an example of an early eosinophilic promyelocyte in the initial stages of prophase. Panel *B*(2) highlights an example of an early, and difficult to identify unambiguously, myeloblast. The following are examples of other cells of interest:

- Panel *A*: (*3*) erythroid series cells
- Panel *B*: (*3*) neutrophilic promyelocyte in early prophase; (*4*) monocyte; (*5*) lymphocyte
- Panel *C*: (*2*) lymphocytes; (*3*) neutrophilic myelocyte; (*4*) orthochromatic erythroblast, (*5*) polychromatic erythroblasts
- Panel *D*: (*4*) lymphocyte
- Panel *E*: (*4*,*5*,*6*) various stages of erythroblasts; (*7*) orthochromatic erythroblast; (*8*) neutrophil promyelocytes

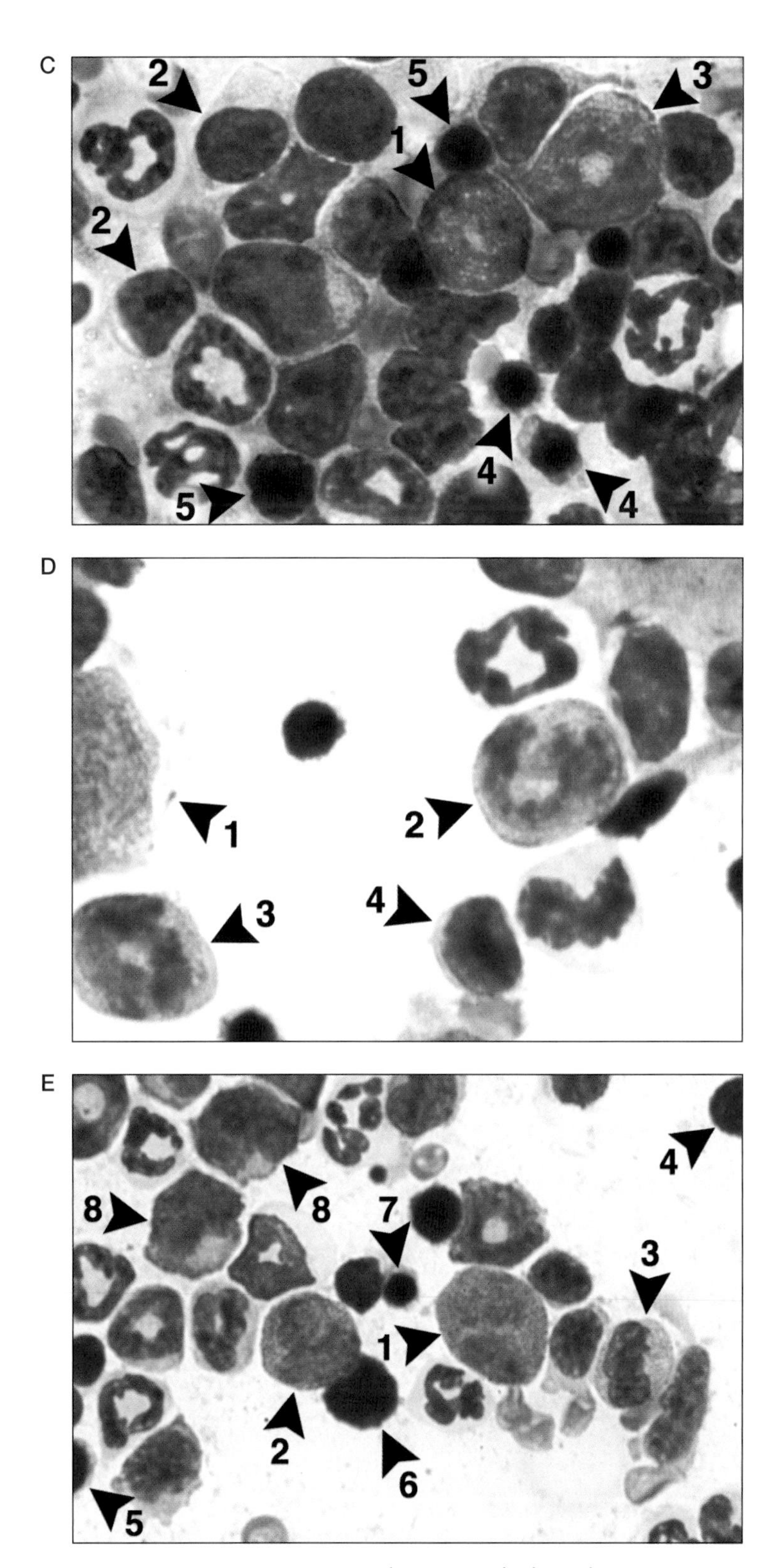

FIGURE 16. *(See facing page for legend.)*

nuclear morphology may present as a "ring" or "figure eight," or the nucleus may be flattened on itself, giving the appearance of a solid cylinder.

Technical Tip: The nuclei of circulating and tissue-dwelling eosinophils display more contorted versions of the "ringed" nuclear structure (e.g., compare the peripheral eosinophils shown in the poster, Mouse Peripheral Blood Cells, with the marrow eosinophils of Fig. 16), with the highest degree of polymorphonuclear character generally displayed by tissue eosinophils such as those in airway exudates (i.e., bronchoalveolar lavage [BAL] eosinophils).

The next earliest stage is the eosinophilic myelocyte. Although similar in many respects to the eosinophilic metamyelocyte, its distinguishing features are that it is capable of at least one terminal replicative event and, perhaps consequently, retains some cytoplasmic basophilia. Eosinophilic myelocytes comprise a substantial fraction of the series within the marrow, and they are easily grouped as those cells with "ringed" nuclei, a nominal degree of heterochromatic condensation, and a basophilic cytoplasm with distinctive eosin-staining granulation that is decidedly muted relative to the most mature cell of this series. It is noteworthy that all three of these morphological features will range considerably. For example, the nuclei of eosinophilic myelocytes may be substantially indented, have only a slight perforation, or be absolutely ring-shaped. The cytoplasmic basophilia of these cells may be "deeply azure" to "pastel," and there may be many granules or only a few. In addition, it is important to keep in mind that eosinophilic myelocytes may be found in various stages of mitosis, adding a distinctive yet recognizable character to their nuclei. As a general rule of thumb, less mature eosinophilic myelocytes will differ from more mature leukocytes (i.e., circulating cells) by having more intense cytoplasmic basophilia (i.e., a blue hue). Moreover, the nuclear-to-cytoplasmic ratio of these cells is distinctly larger than more mature cells, and the folded ring shape of their nuclei is less evident. Thus, a series of decreasingly mature *eosinophils* can be created from the photomicrographs of Figure 16 based on these cytoplasmic/nuclear issues, i.e., Fig. 16, (D)2 and (D)3; Fig. 16, (A)1 and (A)2; Fig. 16, (B)1; Fig. 16, (C)1; and Fig. 16, (E)1 and (E)2, respectively.

Finally, the cell that gives rise to the eosinophilic myelocyte is the eosinophilic promyelocyte. This cell differs from the more mature myelocyte by several subtle features. The nuclear-to-cytoplasmic ratio is higher, the nucleus is not ringed but rather configured with a slight to substantial indentation, and it may contain one to several nucleoli. The cytoplasmic basophilia is also more intense. There are few if any granules, and these are frequently restricted to the area of the nuclear indentation. Late eosinophilic promyelocytes will have a few of the distinct lineage-specific mature eosin-staining (magenta/orange/pink) granules, and the area of the cytoplasm at the nuclear indentation will be unstained and co-located with the Golgi apparatus. As this cell develops, more of the distinct, lineage-specific granules will become apparent. The early eosinophilic promyelocyte is mitotically very active. Consequently, considerable variation in shape, mostly in cross-sectional diameter, will be encountered and, in the absence of definitive granulation, an unambiguous lineage assignment may not be possible.

The earliest cell in the series that may be identified as "myelopoietic" is the eosinophil myeloblast. Unfortunately, this cell is virtually indistinguishable from the earliest cells of the other granulocytic lineages (see Figs. 15,[C]4 and 17,[B]1 and [G]6). The features of myeloblasts include a very high nuclear-to-cytoplasmic ratio, a very azurophilic, although scant, cytoplasm that is relatively homogeneous, and a slight clear area (Golgi apparatus) occupying a small region of cytoplasm adjacent to the nucleus. In addition, the nuclei of these cells are quite spherical, although indented at the site of the Golgi apparatus. Several nucleoli are almost always evident.

Did you know? Relative to their human counterparts, mouse eosinophils stain poorly with eosin for reasons that up to a few years ago were unknown. Instead of a vibrant magenta, which is easily seen in even histopathological assessments of dense human tissue sections, mouse eosinophils stain a pale red color. This phenomenon was understood with the cloning and characterization of the proteins comprising the respective eosinophil secondary granules. These data demonstrated that, unlike the very cationic character of the human granule proteins, the mouse granule proteins were only slightly basic, with pIs closer to 7.0–8.0 (Rothenberg and Hogan 2006).

Neutrophil Granulocytes

Much of what has been said about the morphology of the eosinophil lineage of granulocytes can be said of cells of the *neutrophil* series. Two important distinctions are worth noting: (1) Unlike the prevalent eosinophil secondary granules, which are composed of abundant cationic proteins and thus stain well with eosin (magenta/orange/pink), the prevalent neutrophil-associated granules are composed of pH-neutral constituents and thus are unstained with classical Romanowsky-dye combinations (Fig. 17). (2) The nuclei of mature neutrophils and/or PMNs are considerably more heterochromatic, with one to several constrictions (i.e., lobulations) relative to eosinophil counterparts of similar maturation (see, e.g., Fig. 17, [A]3).

Did you know? A neutrophil from a female mouse often displays a vestigial nuclear lobe (a "drumstick") as a result of an inactivated X chromosome. This "drumstick" is a definitive characteristic of a female-derived PMN (i.e., a chromosomally XX subject animal). An example is clearly visible in the photomicrograph of Figure 21, C(3).

In learning to differentiate the PMN myeloid precursors, it is advisable to review all of the morphological forms presented by cells that are distinctly

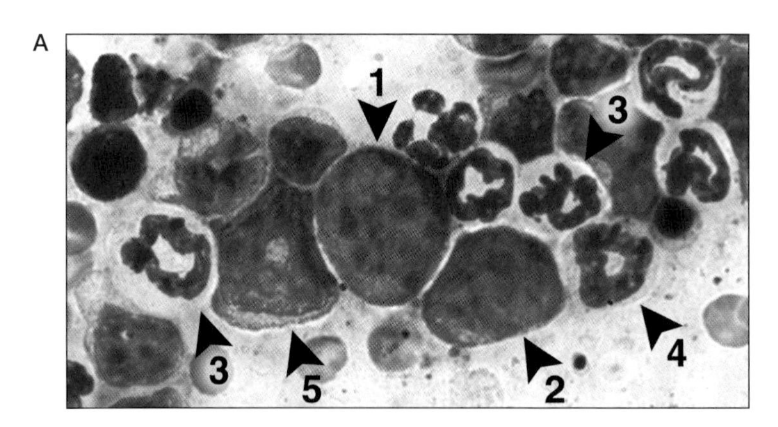

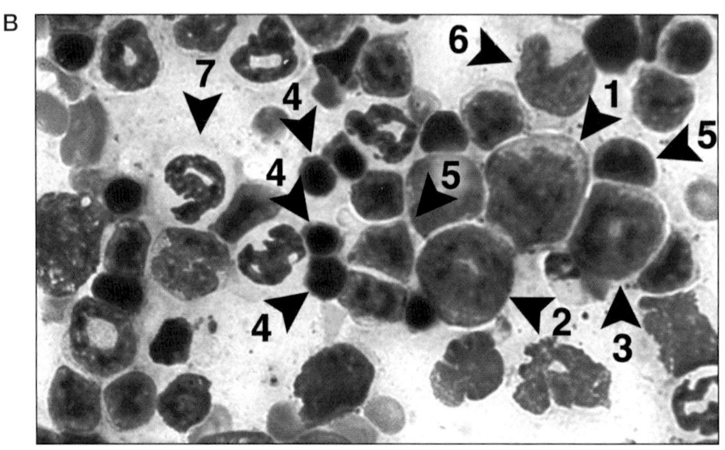

FIGURE 17. Differential assessment of the mouse neutrophil lineage. The neutrophilic myeloid series is the most abundant granulocyte population of normal mouse marrow. Mature neutrophils and/or PMNs (neutrophilic metamyelocytes) are normally found within the marrow as terminally differentiated cells ready for emergence into peripheral circulation. Their nuclei are generally ring-shaped with a definitive degree of chromatin condensation, giving the nuclei an almost lobular appearance. The cytoplasm of these mature neutrophils is virtually clear, often with evidence of unstained granulation (panels *A*[3] and *B*[7]). In contrast, neutrophilic myelocytes exhibit a wide range of morphologies (panels *A*[5], C[2], C[6], *D*[2], *D*[3], *E*[2], *E*[6], *E*[7], and *F*[3]). Specifically, neutrophilic myelocytes have little or no nuclear polylobulation, although the nucleus varies from slightly perforated to ringed, but with no condensation except that which may be related to mitosis. Their cytoplasm also has a range of granule content within a basophilia that varies between pale sky blue and an intense iridescent blue, which allows for a better demarcation of the definitive unstained granules. Neutrophilic promyelocytes will have slight to moderately indented nuclei with cytoplasm that has little definitive granulation (panels *A*[1], *A*[2], C[1], *F*[1], and *F*[2]). Comparison of putative neutrophilic promyelocytes with similarly staged eosinophilic promyelocytes is helpful because the latter have definitive acidophilic (magenta/orange/ pink) granulation at that stage and comparisons with other features is made easier (compare the eosinophilic promyelocytes of Fig. 16, *E*[1] and *E*[2] with the neutrophilic promyelocytes noted in this figure). The following are examples of other cells of interest:

- Panel *A*: (*4*) early neutrophilic myelocyte
- Panel *B*: (*1*) indeterminate myeloblast; (*2,3*) neutrophilic myelocytes; (*4*) erythroblast; (5) lymphocyte; (6) monocyte

C

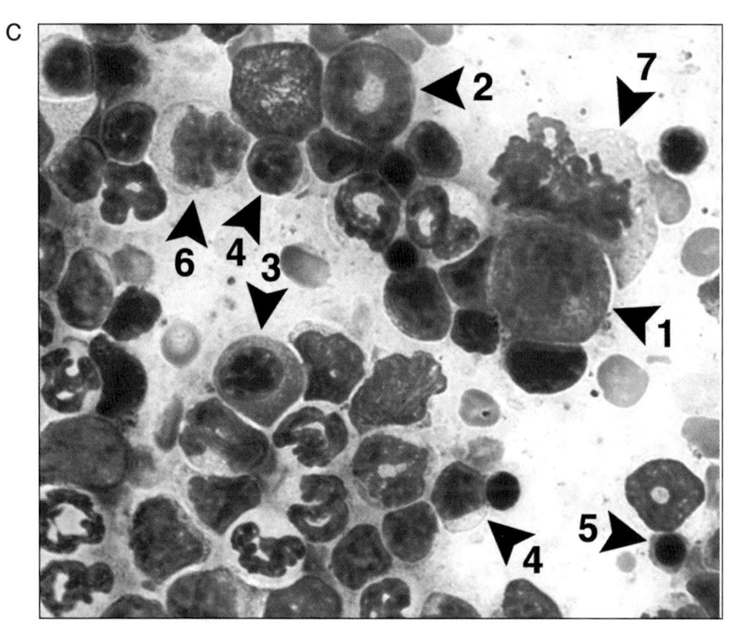

D

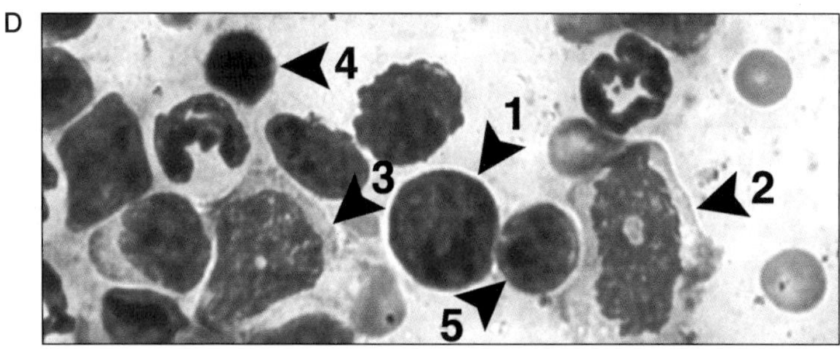

E

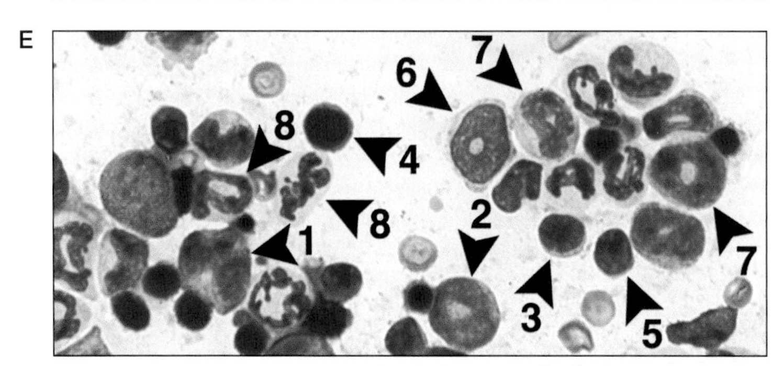

FIGURE 17. *(Continued)*

- Panel *C*: (*3*) proplasma cell; (*4*) lymphocyte; (*5*) polychromatic erythroblast; (*7*) a cell in metaphase, likely in the neutrophil lineage series, although possibly a monocyte
- Panel *D*: (*1*) myeloblast; (*4*) basophilic erythroblast; (*5*) lymphocyte
- Panel *E*: (*1*) late neutrophilic promyelocyte; (*3,5*) lymphocytes; (*4*) erythroblast; (*8*) neutrophilic metamyelocyte

(Continued on following page)

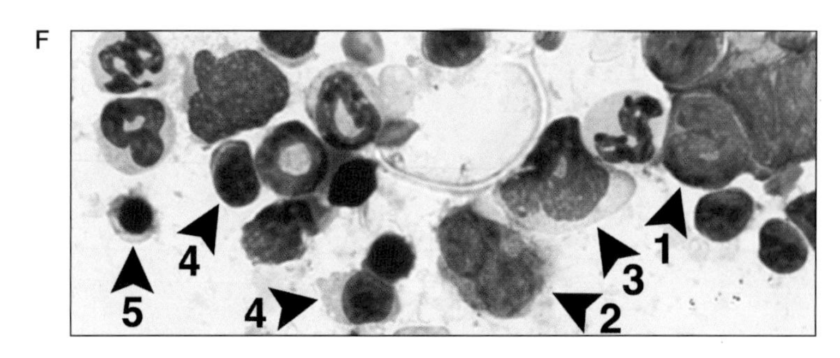

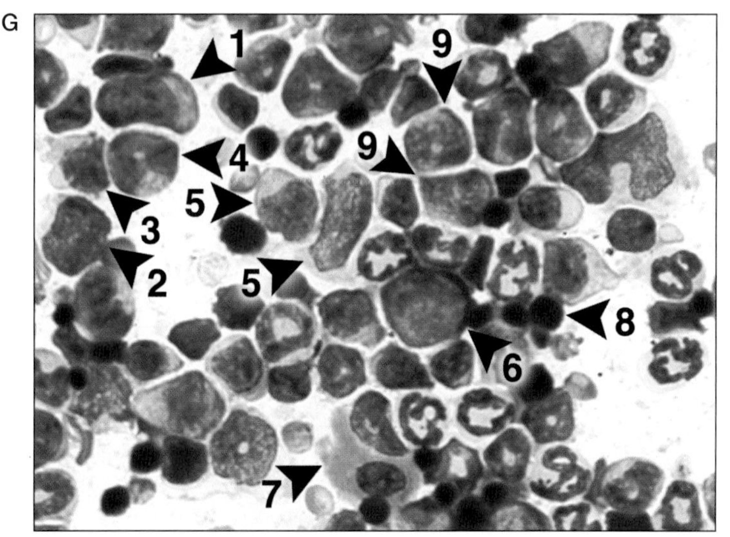

FIGURE 17. *(Continued)*

- Panel *F*: (*4*) lymphocyte; (*5*) orthochromatic erythroblast
- Panel *G*: (*1*) indeterminate promyelocyte; (*2*) undefined; (*3*) orthochromatic erythroblast; (*4*) neutrophilic myelocyte; (*5*) monocytes; (*6*) indeterminate myeloblast; (*7*) plasma cell; (*8*) erythroblast; (*9*) late neutrophilic promyelocyte

eosinophils. It is possible, then, to make relevant comparisons identifying similar cells that do not have the definitive eosin-staining granules; these will most likely be neutrophil-lineage precursors. At every stage, from the mature metamyelocytes back to the least-mature late promyelocytes, maturing cells of the two lineages are comparable except for the granulation and the differences in nuclear condensation. Both proliferate and mature through the same stages, with relatively similar effects on cytoplasmic basophilia (i.e., blue hue) and the level (but not the color!) of the accumulating granules. The nuclear characters of developing cells of both lineages, including the various stages of mitosis and nuclear-to-cytoplasmic ratios, are also quite similar. The single most unique feature of early lineage-committed cells of these granulocytic lineages is that the least-mature PMNs become difficult to distinguish from lymphoid, monocytic, and even erythroid progenitors because the nonstaining neutral cytoplasmic granules of PMNs may be confused with cytoplasmic structures in these other lineages. However, this is not usually the case with the eosinophil precursors because of the distinctive stain coloration (magenta/orange/pink) associated with these granules.

More clues for accurately classifying neutrophils may be obtained by associating the stages of nuclear maturation with those of cytoplasmic maturation. For example, if a cell has pale blue cytoplasm with no evidence of granulation and a low nuclear-to-cytoplasmic ratio, it is *not* an early progenitor. Likewise, intensely blue cytoplasm, a kidney-shaped nucleus, and a low nuclear-to-cytoplasmic ratio are not consistent with a more mature cell morphology. Thus, the defining features of the differentiating lineage must be coordinated. That is, as the leukocytes mature from the earliest progenitor cell, the nuclear-to-cytoplasmic ratio decreases, heterochromatic condensation increases, cytoplasmic basophilia (i.e., blue hue) diminishes, and definitive end-products associated with mature, terminally differentiated cells accumulate.

Basophil Granulocytes

The *basophil* is an extraordinarily rare leukocyte in the mouse and controversy surrounds its identification (Lee and McGarry 2007). Because the probability of encountering these leukocytes is low, only a brief description is provided here. Mouse basophils, similar to their human counterparts, display specific characteristics that are unique to this lineage. They are similar in size to other blood granulocytes (i.e., eosinophils and neutrophils); however, they contain intensely metachromatic-staining (i.e., very dark blue or almost black) granules that are asymmetrically distributed throughout the cytoplasm and lobulate nuclei that have comparatively weakly staining chromatin (see the poster, Mouse Peripheral Blood Cells).

It is noteworthy, however, that the existing mouse literature describes a leukocyte of low prevalence in the mouse (~1% of marrow cells) that is reported to have a surface immune phenoytpe consistent with a basophil (i.e., FSC^{lo}/SSC^{lo}, Thy-1.2$^+$, CD69$^+$, CD49b(DX-5)$^+$, FcεRI$^+$, 2B4$^+$, CD11b^{dull}, CD24$^-$, CD19$^-$, CD80$^-$, CD14$^-$, CD23$^-$, Ly49C$^-$, CD122$^-$, CD11c$^-$, Gr-1$^-$, α_4 and β_7-integrin$^-$, c-Kit$^-$, NK1.1$^-$, CD3$^-$, B220$^-$, $\gamma\delta$TCR$^-$, $\alpha\beta$TCR$^-$), yet displays a very distinct and uncharacteristic morphology following staining with a Romanowsky-dye combinations (see, e.g., Voehringer et al. 2004). Specifically, these cells display nuclei that are polymorphic and cytoplasmic regions that are slightly basophilic and remarkably devoid of granulation (i.e., stained as well as unstained granules). Clearly, issues surrounding the identification of basophils in the mouse remain, and their resolution is critical before data from human patients and mouse models of disease can be integrated.

Technical Tip: Mast cells are tissue leukocytes that function in immune-mediated inflammation and are marrow derived, although they are not normally evident in marrow as mature cells. They derive their name from the German word "masten," meaning fattened or well fed. Paul Erlich applied this name to these cells because their cytoplasm is "glutted" with darkly blue-to-black-staining granules, often to the point of obscuring the nucleus and other cytoplasmic details. It is very easy to mistakenly identify mast cells as basophils. However, it is important to keep in mind that terminally differentiated mast cells are, on average, three to five times larger than marrow-derived leukocytes (with the exception of megakaryocytes). Mast cells are not found in the marrow of otherwise healthy mice, and they are never found in peripheral circulation. That is, unlike circulating granulocytes, mast cells mature and reside exclusively in the outlying tissues, although they are derived from stem cells of marrow origin (Kitamura et al. 1981; Hayashi et al. 1983).

LYMPHOMONONUCLEAR SERIES

Lymphocytes

Lymphocytes in the bone marrow of the mouse are essentially identical to the mature lymphocytes that appear in the peripheral circulation, spleen, and lymph nodes (Fig. 18). Generally, they are small, nucleated cells with scant cytoplasm. Typically, they are slightly smaller than eosinophils/neutrophils, but "medium" lymphocytes—larger versions of the same cell—are occasionally encountered (<10% of all lymphocytes). A lymphocyte is not a difficult leukocyte to identify. Its major "conflicters" are the basophilic erythroblast and early forms of the other mononuclear lineages (i.e., the monocytic and erythropoietic lineages). Lymphocytes are distinguished from erythroid cells by two main features: Their nuclei are less coarse (i.e., less textured) and the cytoplasm is not as intensely basophilic. They are different from monocyte progenitors in that the nuclei of lymphocytes are more spheroid (usually no hints of indentation) and decidedly asymmetrically positioned in the cell. In addition, the cytoplasm of lymphocytes is even more "pale blue" relative to monocytes and their progenitors, and in general, lymphocytes are slightly smaller. That being said, confusion is still best resolved by comparing the cell in question to other lymphocytes in adjacent fields that are more definitive of the lineage. The prolymphocytes and lymphoblasts are truly rare and virtually indistinguishable from the blast cells of other lineages.

Technical Tip: Any marrow preparation that has a predominance of what might be considered "lymphoblasts" should be considered abnormal, and special approaches using immunophenotyping by immunocytochemistry and/or FACS analysis should be used to confirm their identity.

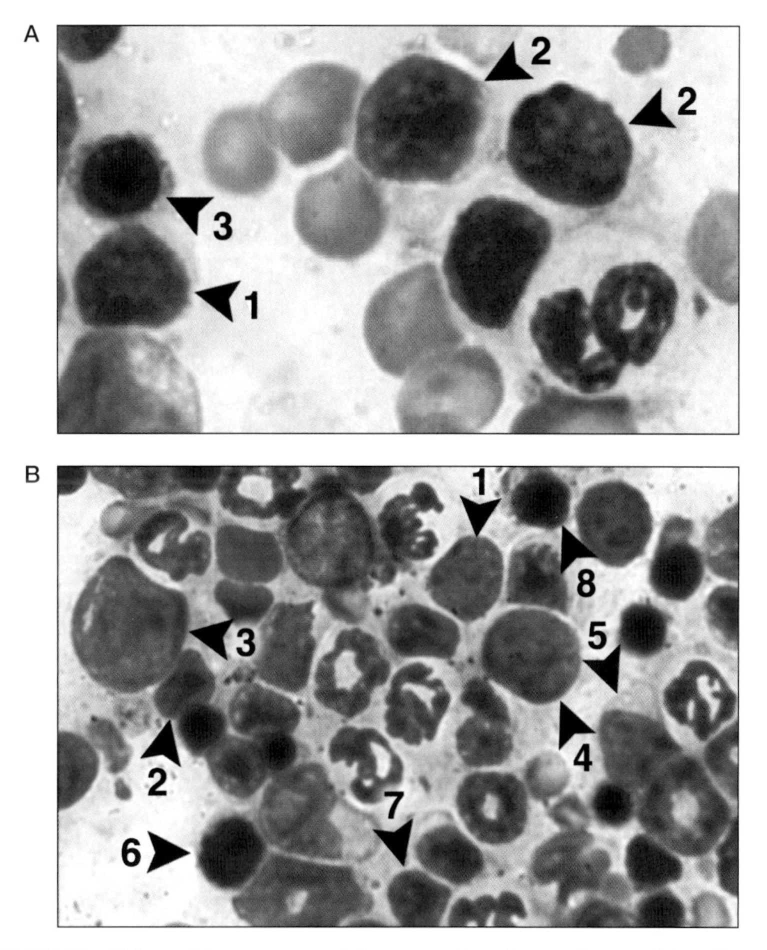

FIGURE 18. Differential assessment of the mouse lymphocyte lineage. The primary confounding cells for lymphocytes within the marrow are erythroid progenitors and monocytes. However, both have features in detail that distinguish them from lymphocytes. A good comparison of a lymphocyte closely resembling a polychromatic erythroblast is found in panel *A*(1) and *A*(2)–lymphoid cells versus panel *A*(3)–erythroid cell. Lymphocytes are slightly larger with paler blue cytoplasm relative to erythroblasts and nuclei that are not as dense. Panel *B*(1,2,5,7)–lymphoid cells similarly may be compared with panel *B*(4)–erythroblast, panel *B*(6)–polychromatic erythroblast, and panel *B*(8)–orthochromatic erythroblast. Panel *B*(3) is a representative photomicrograph of a myeloblast with nucleoli. The slight deeply blue cytoplasm, nucleus with nucleoli, and a slight clear area (Golgi apparatus) adjacent to the nucleus indicate this designation. Its distinction from panel *B*(4)–erythroblast (i.e., the absence of an apparent Golgi) should be noted.

Plasma Cells (Ig-secreting B Cells)

A *plasma cell* (Figs. 14, B and 19) is a variant of the lymphocyte whose prevalence in marrow and circulation requires an investigator to be aware of its existence and to identify it with confidence. A plasma cell has the same general morphology as a typical lymphocyte, and two specific features are unique to these cells: (1) Plasma cells have intensely basophilic

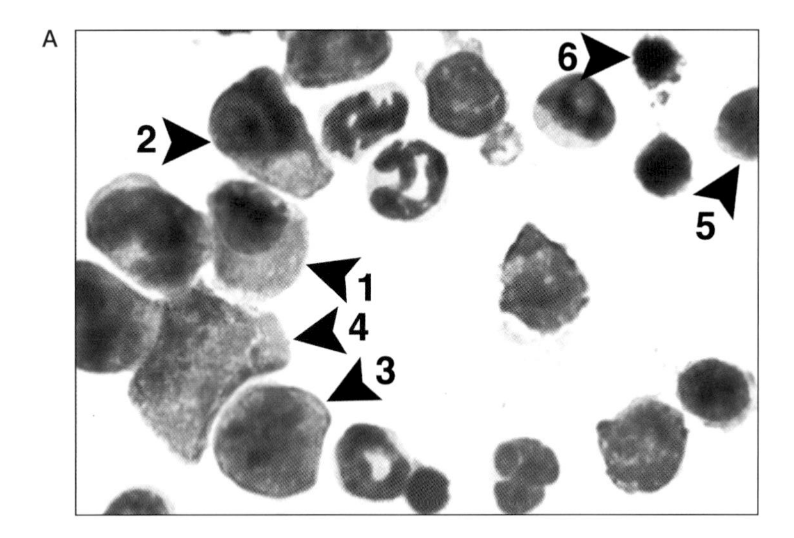

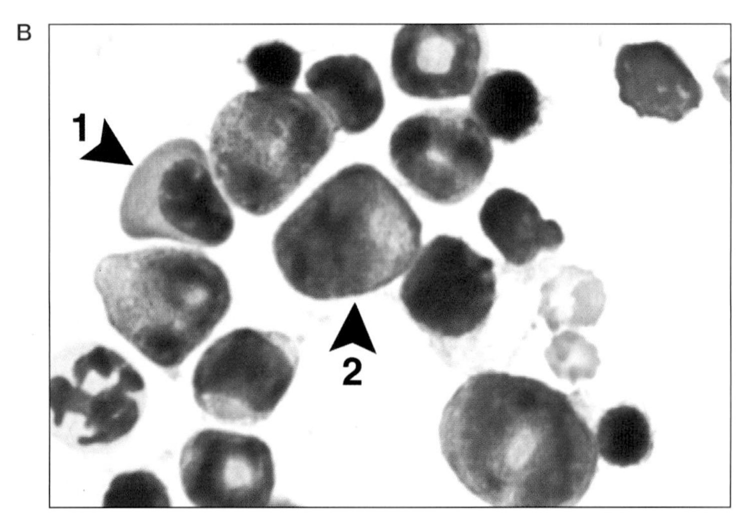

FIGURE 19. Differential assessment of the mouse plasma cell lineage. Plasma cells are particularly identifiable from early stages by virtue of the definitive eccentric nucleus with an adjacent clear area (Golgi apparatus) within a relatively homogenous blue (basophilic) cytoplasm. It is noteworthy that the nuclear-to-cytoplasmic ratio decreases with maturity of these cells, as is true for most hematopoietic cell lineages, i.e., the proportion of the cell occupied by the nucleus is much greater, approaching ~90% in the early plasma cell, and decreases progressively to 20%–40% in mature plasma cells. Early plasma cells are evident in panels *A*(1), A(2), *B*(1), *B*(2), and *C*(1). In contrast, panels *D*(1) and *E*(1) are representative examples of more mature plasma cells. The resemblance of other cells within the field to plasma cells is reflective of the confounding members of other lineages, most notably erythroid and lymphoid series cells. However, the unique (i.e., specific) characteristics of mature plasma cells are readily identifiable even within a complex marrow smear. The following are examples of other cells of interest:

- Panel *A*: (3) myeloblast with three nucleoli; (4) early indeterminate promyelocyte; (5) lymphocyte; (6) polychromatic erythrocyte
- Panel C: (2) eosinophilic myelocyte; (3) neutrophilic myelocyte; (4) neutrophilic metamyelocyte
- Panel *D*: (3) lymphocyte; (4) erythroblast; (2,5,6) monocytes
- Panel *E*: (2) early eosinophilic myelocyte

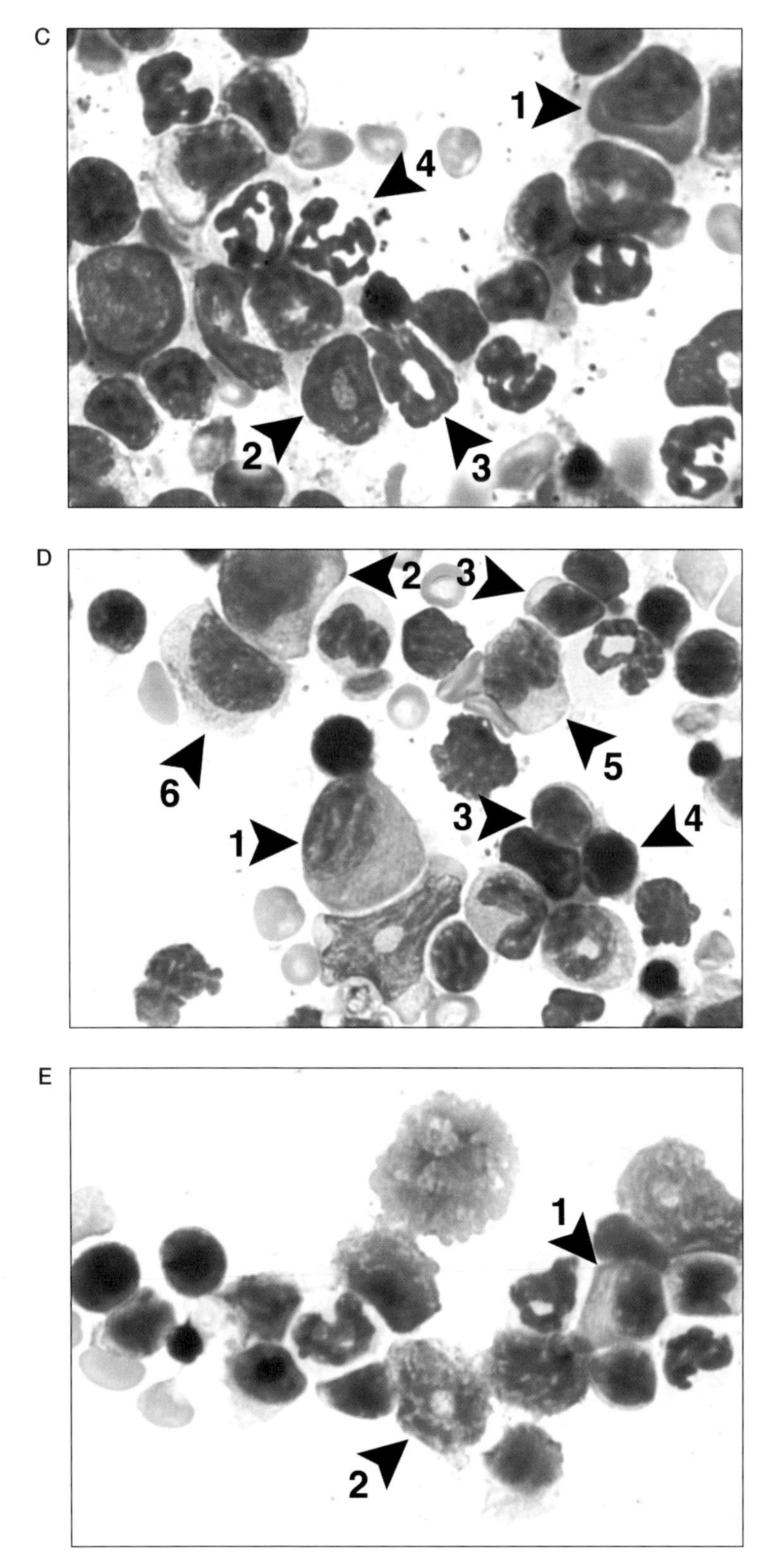

FIGURE 19. *(See facing page for legend.)*

(i.e., blue) cytoplasm and (2) the cytoplasm is essentially featureless (i.e., no hints of granulation) with a clear area (an extensive Golgi) adjacent to an eccentric and well-demarcated nucleus. Distinguishable stages in the differentiation of this unique lymphocytic lineage will help in the identification of these cells relative to other mononuclear leukocytes. It is unusual to confuse it with any other lineage because of these unique morphological characteristics. The mature plasma cell is an active antibody-secreting B cell. The nuclear-to-cytoplasmic ratio of plasma cells is somewhat smaller than that of other lymphocytes, owing to a larger cytoplasmic area associated with antibody production (compare the photomicrographic images in Fig. 18 with those in Fig. 19). Mature plasma cells are not generally mitotic in the marrow under homeostatic baseline conditions.

The precursor to the mature plasma cell is the *proplasma cell* (or the *pro-B cell*). It differs from the mature form by having a more intensely basophilic cytoplasm, a lesser developed Golgi, and thus a higher nuclear-to-cytoplasmic ratio. The spherical nuclei of mitotically active plasma cells may be slightly less eccentric and smaller than their more mature, nondividing progeny. The least-mature identifiable cell of the series is the *plasmablast*. Similar to other very early cells, it is best identified in the context of slightly more mature cells of the same series. Despite these difficulties, it is still possible, with experience and effort, to retrace the developmental pathway of these leukocytes to less mature cells until the progenitors are indistinguishable from those of other lymphoid, monocytic, and even erythroid lineages.

Did you know? Unlike other WBCs, which develop and mature in the bone marrow before their exit into peripheral circulation, B and T lymphocytes leave the marrow as immature "pro-" forms and complete maturation in outlying (i.e., extramedullary) lymphoid tissues. Once achieving a differentiated state, the mature cells—the B and T lymphocytes characteristic of peripheral blood—reenter circulation. Indeed, the names of these lymphocytes reflect this unique route of differentiation: B cells are named after the bursa of Fabricius, the location in chickens from which the cells are derived (Sato and Glick 1970). T cells are so named because the thymus is the target organ of pro-T lymphocytes upon exit of the marrow (Sell and Max 2001).

Monocytes

Mature cells of the monocyte lineage in the marrow are similar to the terminally differentiated monocytes observed in peripheral blood (Fig. 20). Nonetheless, cells of the monocytic lineage are often easily confused with lymphocytes. Proficiency in the accurate identification of monocytes comes with experience and practice. Four important distinctions allow an investigator to distinguish these very similar looking cells: (1) Unlike the distinctly spherical nucleus of a cell in the lymphoid series, the nuclei of

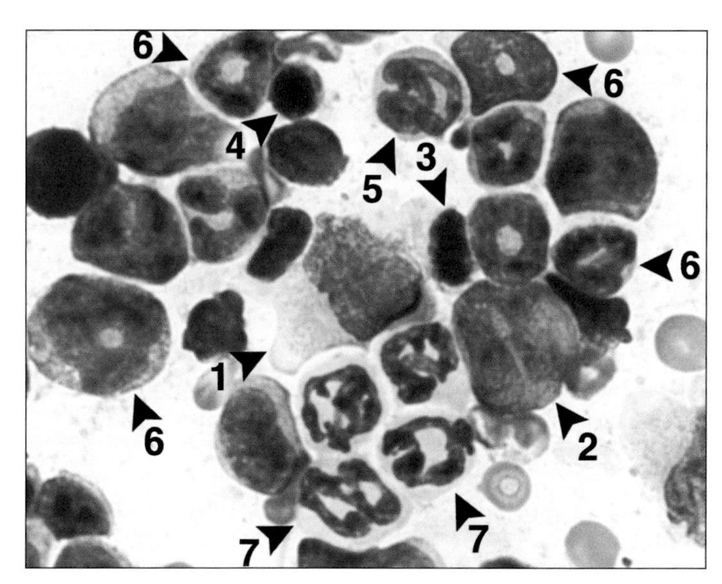

FIGURE 20. Differential assessment of the mouse monocyte lineage. An abundance of pale blue cytoplasm, a nuclear-to-cytoplasmic ratio lower than that typical of lymphoid series cells, a relatively immature nucleus (often with visible nucleoli), and no cytoplasmic granulation all typify monocytic lineage cells (*1*). These cells may be compared with other cells within the field of this figure, such as an early eosinophil myelocyte (*2*), several neutrophils (unmarked) of varying levels of maturity, and orthochromatic (*3*) and polychromatic (*4*) erythroblasts. Photomicrographs of other representative examples of monocytes are found in Fig. 17, *B*(6) and *C*(5). The following are examples of other cells of interest: (5) late eosinophil myelocyte, (6) neutrophil myelocyte, and (7) neutrophil metamyelocytes.

monocytic cells have a characteristic "kidney shape" and/or a slight indentation, giving them a bilobed appearance. (2) The nuclei of monocyte lineage cells are usually centered within the cell so that cytoplasm (even if it is a scant amount) is clearly identifiable around the entire nucleus. (3) The nuclear-to-cytoplasmic ratio of monocyte lineage cells is smaller than that of equivalent lymphoid cells. (4) The cytoplasm of monocyte lineage cells is pale blue-gray and thus is generally less basophilic relative to equivalent cells in the lymphocyte series. An additional, although less reliable, feature that distinguishes monocyte lineage cells is the observation that some inclusion bodies, such as mitochondria and vacuoles, may occupy the cytoplasm.

Technical Tip: Aside from the difficult task of differentially identifying monocytes relative to leukocytes in the lymphoid series, it is helpful to consider the identification of monocyte cells in the context of lineage cell comparisons from three specific series: monocytes versus eosinophils, monocytes versus neutrophils, and finally eosinophils versus neutrophils. The cells of the eosinophil lineage present a clear spectrum of morphologies characteristic of "granule-poiesis," and neutrophils display a unique spectrum of polymorphonuclear changes in the context of increasingly abundant unstained granules. Once appreciated, the differentiation of the PMN series from cells of the monocyte lineage becomes easier. Recognize, however, that virtually all cells comprising the "blast" compartment are very difficult to classify by unique lineage.

Less mature cells of this lineage have a slightly more spherical nucleus, a more basophilic (i.e., blue hue) cytoplasm, and an irregular shape overall. The more primitive cells (monoblasts and promonocytes) are not easily identified. Subtle differences may appear to exist between the progenitors of monocytes and those of other lineages, especially with regard to nuclear and cytoplasmic "texture," but these differences likely reflect impressions more than objective criteria. The very early stages, however, comprise a very low proportion of the poietic population. The increased prevalence of monoblasts and promonocytes in the marrow is generally associated with instances of a peripheral monocytosis, in which there would be a concomitant increase in more mature forms, or in mononuclear cell leukemia, which is evidenced by a disproportionate increase in immature cells of this lineage.

Megakaryocyte

The *megakaryocyte* (Fig. 21) is perhaps the most easily distinguished cell in the marrow because it is unique to this site (and other sites with hematopoietic activity such as the spleen) and is several times the size of any of its neighbors—often as much as five times! It has a single very large nucleus, owing to repeated endonuclear duplication (i.e., karyokinesis without cytokinesis); this distinguishes it from multinuclear giant cells of inflammation (Warren 1980). Megakaryocytes have bulky excessive cytoplasm with a blue cast. Because these large cells are particularly susceptible to

FIGURE 21. Differential assessment of the mouse megakaryocyte lineage. The development of distinctive megakaryocytes progresses from precursors to cells that are considerably larger than their neighbors. Intermediate forms may have a center perforation in the nucleus (panel *A*[1]). Even amidst other marrow cells, they are unmistakable regardless of their relative state of maturation (panel *A*[1] vs. *B*[1] vs. *C*[1,2]). Brush smears, as opposed to cytospin preparations, are especially preferred if the structural morphology of megakaryocytes is to be maintained. Many megakaryocytes (panels *A* and *C*) have cells of other lineages within their cytoplasm. This feature is termed emperipolesis and is not considered reflective of any pathology. The following are examples of other cells of interest:

- Panel *A*: (2) eosinophilic myelocyte; (3) neutrophilic metamyelocyte
- Panel *B*: (2) late myeloblasts; (3) early eosinophilic myelocyte
- Panel *C*: (2) early megakaryocyte; (3) neutrophilic metamyelocyte with "drumstick" nuclear appendage

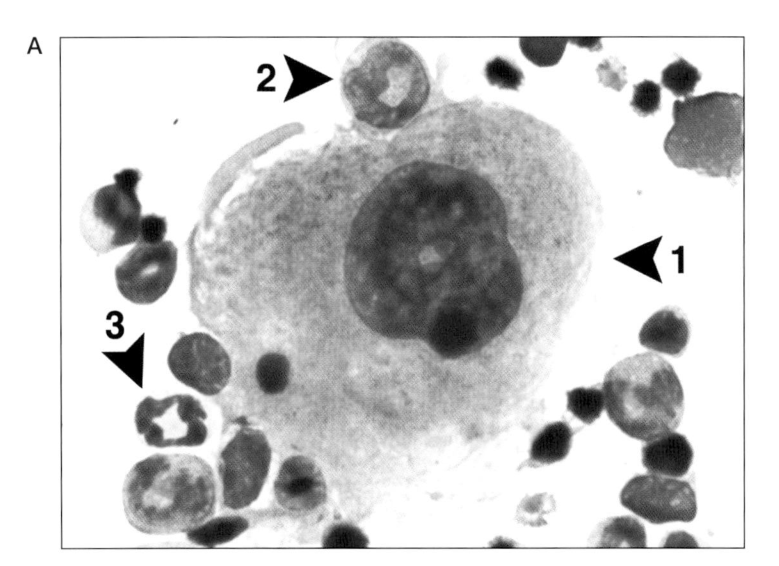

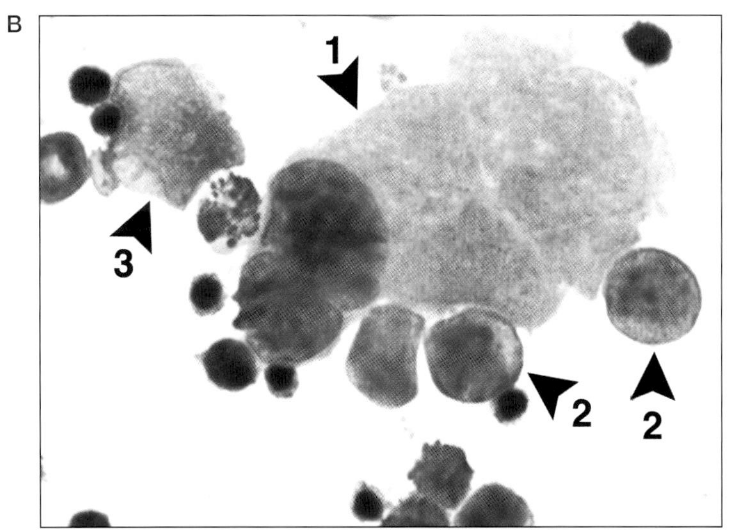

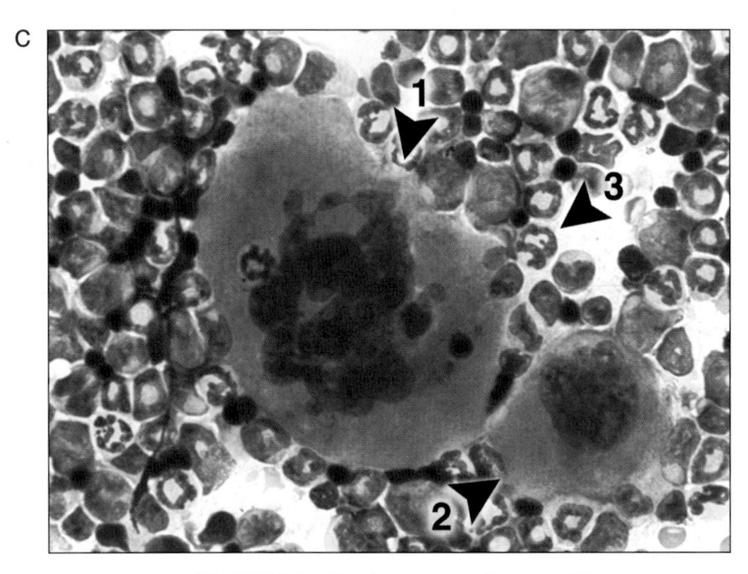

FIGURE 21. *(See facing page for legend.)*

sheer forces, they are preferentially preserved on slides using the brush smear technique (see Protocol 10), which is gentler than the cytospin method (see Protocol 11). The behemoth character of this marrow cell, including its increased nuclear content and copious cytoplasm, is by current understanding preparatory for the remarkable production of platelets, which are extruded as cytoplasmic fragments from these cells by a process known as "budding" (Michelson 2007).

Technical Tip: As much as 50% of the megakaryocytes in the marrow internalize other intact cells, and this will be evident within the cytoplasm. The role and/or importance of this phenomenon, termed emperipolesis, is unknown but nonetheless creates a striking image (see Fig. 21).

The chromatin content of megakaryocytic lineage cells increases from 2N in the hematopoietic stem cell to 32N–64N in the mature megakaryocyte. As a consequence, the overall size of the cell consistently increases with maturity. Rarely, multipolar anaphases may also be encountered that are visually spectacular! Indeed, this is a relatively accurate, if not quick, assessment of lineage development. Regardless of the level of maturity, once the nucleus acquires >8N DNA, as evident by its size, it is conspicuously a megakaryocyte. Even early megakaryocytes are slightly larger than neighboring leukocytes and have nuclei that are easily recognized as belonging to the lineage. As with several of the other lineages, an excess of megakaryocytes should be considered evidence of an underlying disorder, perhaps related to modest blood loss.

BLAST CELLS AND OTHER EARLY PROGENITORS

Some cells within the marrow population defy classification (see Figs. 15,[C]4; Fig. 17,[B]1, [G]2, and [G]6; and 19,[A]3). Although the destiny of these blast cells and early progenitors is to produce one or more of the identified cell lineages of blood, it is not possible to determine which pathway they will follow. Lineage commitment may not have occurred in these cells, and it is not the intention of this overview to offer suggestions regarding the subtle differences among the most immature cells of hematopoietic populations. Suffice it to say that these cells comprise a very small component of the population (<0.1% of total marrow cells). As such, it is neither practical nor instructive to consider descriptions of these cells at length. However, the presence of significant numbers of these cells in the marrow would be an indication of significant pathology, including disease and/or trauma.

Conclusion

THE PROCEDURES, DESCRIPTIONS, FIGURES, and tables presented herein constitute only a beginning. They should serve as a primer for the methodical unraveling of the complex cellular morphologies displayed by the hematopoietic tissues and peripheral blood of mice. The photomicrographs of marrow and blood cells that were selected for presentation are typical of what is encountered in the process of assessment by an investigator using a mouse model of human disease. To this end, they should serve as both guide and reference to develop the skills and experience needed to identify the formed elements of the blood of the laboratory mouse. The guidelines for assessment presented here begin with identification of the most typical mature cells of each lineage. From this starting point, it becomes less challenging to correctly identify the composite cells of the blood, marrow, and other hematopoietically active tissues in mice.

Final Technical Tip: It is important to realize that morphological differences exist between developing and mature leukocytes of the various and ever-increasing numbers of mouse strains available to investigators. Sometimes, subtle, slight differences in the staining qualities of these cells may lead to the misinterpretation of results.

There is no absolute set of criteria to establish classifications. Consistency is critical, and eventually, the necessary proficiency will be developed to permit the confident assessment of hematopoiesis. Moreover, there is no substitute for the commitment of time required for the microscopic examination of peripheral blood and marrow. However, the rewards are enormous! In addition to the successful maintenance of a healthy mouse colony, the rapid and inexpensive assessments described here can save valuable time and resources essential to the design, development, and interpretation of experiments involving the pathological changes associated with mouse models of human disease, as well as other phenomena affecting the lymphohematopoietic system.

References

Christensen SD, Mikkelsen LF, Fels JJ, Bodvarsdottir TB, Hansen AK. 2009. Quality of plasma sampled by different methods for multiple blood sampling in mice. *Lab Anim* **43:** 65–71.

Custer RP, Hayhoe FGJ. 1974. *An atlas of the blood and bone marrow,* 2nd ed. Saunders, Philadelphia.

Golde WT, Gollobin P, Rodriguez LL. 2005. A rapid, simple, and humane method for submandibular bleeding of mice using a lancet. *Lab Anim* **34:** 39–43.

Hayashi C, Sonoda T, Kitamura Y. 1983. Bone marrow origin of mast cell precursors in mesenteric lymph nodes of mice. *Exp Hematol* **11:** 772–778.

Hoff J. 2000. Methods of blood collection in the mouse. *Lab Anim* **29:** 47–53.

Kitamura Y, Yokoyama M, Matsuda H, Ohno T, Mori KJ. 1981. Spleen colony-forming cell as common precursor for tissue mast cells and granulocytes. *Nature* **291:** 159–160.

Koury MJ, Mahmud N, Rhodes MM. 2009. Origin and development of blood cells. In *Wintrobe's clinical hematology*, 12th ed. (ed. JP Greer et al.), Vol. 1, pp. 79–106. Lippincott Williams & Wilkins, Philadelphia.

Lee JJ, McGarry MP. 2007. When is a mouse basophil not a basophil? *Blood* **109:** 859–861.

Lillie RD. 1977. *H.J. Conn's Biological stains: A handbook on the nature and uses of the dyes employed in the biological laboratory*, 9th ed. Williams & Wilkins, Baltimore.

McClung CE (McClung Jones R, ed.) 1961. *Handbook of microscopical technique for workers in animal and plant tissues,* 3rd ed. Hafner, New York.

McGarry MP, Stewart CC. 1991. Murine eosinophil granulocytes bind the murine macrophage-monocyte specific monoclonal antibody F4/80. *J Leukoc Biol* **50:** 471–478.

Michelson AD. 2007. *Platelets*, 2nd ed. Academic Press, Amsterdam.

Morton DB, Abbot D, Barclay R, Close BS, Eubank R, Gask D, Heath M, Mattic S, Poole T, Seamer J, et al. 1993. Removal of blood from laboratory mammals and birds. First report of the BVA/Frame/RSPCA/UFAW Joint Working Group on Refinement. *Lab Anim* **27:** 1–22.

Rothenberg ME, Hogan SP. 2006. The eosinophil. *Annu Rev Immunol* **24:** 147–174.

Russell ES, Bernstein SE. 1975. Blood and blood formation. In *Biology of the laboratory mouse*, 2nd ed. (ed. EL Green and Roscoe B. Jackson Memorial Laboratory), pp. 351–372. Dover, New York.

Sato K, Glick B. 1970. Antibody and cell mediated immunity in corticosteroid-treated chicks. *Poult Sci* **49:** 982–986.

Schalm OW, Jain NC, Carroll EJ. 1975. *Veterinary hematology,* 3rd ed. Lea & Febiger, Philadelphia.

Sell, S, Max EE. 2001. *Immunology, immunopathology, and immunity,* 6th ed. ASM Press, Washington, DC.

Tavassoli M, Yoffey JM. 1983. *Bone marrow, structure and function*. A.R. Liss, New York.

Voehringer D, Shinkai K, Locksley RM. 2004. Type 2 immunity reflects orchestrated recruitment of cells committed to Il-4 production. *Immunity* **20:** 267–277.

Warren KS. 1980. The cell biology of granulomas (aggregates of inflammatory cells) with a note on giant cells. In *The cell biology of inflammation* (ed. G Weissmann), pp. 543–557. Elsevier, Amsterdam.

Cautions

PLEASE NOTE THAT THE CAUTIONS APPENDIX in this manual is not exhaustive. Readers should always consult individual manufacturers and other resources for current and specific product information. Chemicals and other materials discussed in text sections are not identified by the icon <!> used to indicate hazardous materials in the protocols. However, without special handling, these materials may be hazardous to the user. Please consult your local safety office or the manufacturer's safety guidelines for further information.

The following general cautions should always be observed.

- **Before beginning the procedure,** become completely familiar with the properties of substances to be used.
- **The absence of a warning** does not necessarily mean that the material is safe, because information may not always be complete or available.
- **If exposed** to toxic substances, contact your local safety office immediately for instructions.
- **Use proper disposal procedures** for all chemical and biological waste.
- **For specific guidelines on appropriate gloves to use,** consult your local safety office.
- **Never pipette** solutions using mouth suction. This method is not sterile and can be dangerous. Always use a pipette aid or bulb.
- **The use of microwave ovens and autoclaves** in the lab requires certain precautions. Accidents have occurred involving their use. If the screw top is not completely removed and there is inadequate space for the steam to vent, the bottles can explode and cause severe injury when the containers are removed from the microwave or autoclave. Always completely remove bottle caps before microwaving or autoclaving.
- **Use extreme caution when handling cutting devices,** such as scalpels, razor blades, or needles. If unfamiliar with their use, have an experienced user demonstrate proper procedures. For proper disposal, use the "sharps" disposal container in your lab. Discard used needles *unshielded,*

with the syringe still attached. This prevents injuries and possible infections when manipulating used needles because many accidents occur while trying to replace the needle shield. Injuries may also be caused by broken Pasteur pipettes, coverslips, or slides.

- **Procedures for the humane treatment of animals** must be observed at all times. Consult your local animal facility for guidelines. Animals are known to induce allergies that can increase in intensity with repeated exposure. Always wear a lab coat and gloves when handling these animals. If allergies to dander or saliva are known, wear a mask.

HAZARDOUS MATERIALS

The following items are specifically flagged with <!> in protocols in this book. In general, proprietary materials are not listed here. Kits and other commercial items as well as most anesthetics, dyes, fixatives, and stains are also not included. Follow the manufacturer's safety guidelines that accompany these products.

Alconox detergent is an irritant and may be harmful by inhalation, ingestion, or skin absorption. Wear appropriate gloves and safety glasses.

Chloroform, $CHCl_3$, is irritating to the skin, eyes, mucous membranes, and respiratory tract. It is a carcinogen and may damage the liver and kidneys. It is also volatile. Avoid breathing the vapors. Wear appropriate gloves and safety glasses and always use in a chemical fume hood.

Crystal Violet can cause severe burns. It may be harmful by inhalation, ingestion, and skin absorption. Wear appropriate gloves and safety goggles and use in a chemical fume hood. Do not breathe the dust.

Ethanol (EtOH), CH_3CH_2OH, is highly flammable and may be harmful by inhalation, ingestion, or skin absorption. Wear appropriate gloves and safety glasses. Keep away from heat, sparks, and open flame.

Ethyl alcohol, *see* **Ethanol**

Formaldehyde, HCHO, is highly toxic and volatile. It is also a possible carcinogen. It is readily absorbed through the skin and is irritating or destructive to the skin, eyes, mucous membranes, and upper respiratory tract. Avoid breathing the vapors. Wear appropriate gloves and safety glasses and always use in a chemical fume hood. Keep away from heat, sparks, and open flame.

Formol, *see* **Formaldehyde**

HCl, *see* **Hydrochloric acid**

Heparin is an irritant and may act as an anticoagulant subcutaneously or intravenously. It may be harmful by inhalation, ingestion, or skin absorption. Wear appropriate gloves and safety glasses.

Hydrochloric acid, HCl, is volatile and may be fatal if inhaled, ingested, or absorbed through the skin. It is extremely destructive to mucous membranes, upper respiratory tract, eyes, and skin. Wear appropriate gloves and safety glasses and use with great care in a chemical fume hood. Wear goggles when handling large quantities.

Isopropanol is flammable and irritating. It may be harmful by inhalation, ingestion, or skin absorption. Wear appropriate gloves and safety glasses. Do not breathe the vapor. Keep away from heat, sparks, and open flame.

Isopropyl alcohol, *see* **Isopropanol**

Methanol, MeOH or **H_3COH,** is toxic, can cause blindness, and is highly flammable. It may be harmful by inhalation, ingestion, or skin absorption. Adequate ventilation is necessary to limit exposure to vapors. Avoid inhaling these vapors. Wear appropriate gloves and safety goggles and use only in a chemical fume hood.

Permount, *see* **Toluene**

Silver nitrate, $AgNO_3$, is a strong oxidizing agent and should be handled with care. It may be harmful by inhalation, ingestion, or skin absorption. Avoid contact with skin. Wear appropriate gloves and safety glasses. It can cause explosions upon contact with other materials.

Toluene, $C_6H_5CH_3$, vapors are irritating to the eyes, skin, mucous membranes, and upper respiratory tract. Toluene can exert harmful effects by inhalation, ingestion, or skin absorption. Do not inhale the vapors. Wear appropriate gloves and safety glasses and use in a chemical fume hood. Toluene is extremely flammable; keep away from heat, sparks, and open flame.

Trypan blue may be a carcinogen and may be harmful by inhalation, ingestion, or skin absorption. Do not breathe the dust. Wear appropriate gloves and safety glasses.

Xylene is flammable and may be narcotic at high concentrations. It may be harmful by inhalation, ingestion, or skin absorption. Wear appropriate gloves and safety glasses and use only in a chemical fume hood. Keep away from heat, sparks, and open flame.

Index

Page numbers ending in f, t, v, or p indicate a figure, table, video, or poster, respectively.